T0259683

SpringerBriefs in Biochemistry and Molecular Biology

More information about this series at http://www.springer.com/series/10196

Wei Xiong • Zhigang Xu

Editors

Mechanotransduction of the Hair Cell

 Springer

Editors
Wei Xiong
School of Life Sciences, IDG/McGovern
Institute for Brain Research
Tsinghua University
Beijing, China

Zhigang Xu
School of Life Sciences
Shandong University
Jinan, Shandong, China

ISSN 2211-9353 ISSN 2211-9361 (electronic)
SpringerBriefs in Biochemistry and Molecular Biology
ISBN 978-981-10-8556-7 ISBN 978-981-10-8557-4 (eBook)
https://doi.org/10.1007/978-981-10-8557-4

Library of Congress Control Number: 2018935907

© The Author(s) 2018
This work is subject to copyright. All rights are reserved by the Publisher, whether the whole or part of the material is concerned, specifically the rights of translation, reprinting, reuse of illustrations, recitation, broadcasting, reproduction on microfilms or in any other physical way, and transmission or information storage and retrieval, electronic adaptation, computer software, or by similar or dissimilar methodology now known or hereafter developed.
The use of general descriptive names, registered names, trademarks, service marks, etc. in this publication does not imply, even in the absence of a specific statement, that such names are exempt from the relevant protective laws and regulations and therefore free for general use.
The publisher, the authors and the editors are safe to assume that the advice and information in this book are believed to be true and accurate at the date of publication. Neither the publisher nor the authors or the editors give a warranty, express or implied, with respect to the material contained herein or for any errors or omissions that may have been made. The publisher remains neutral with regard to jurisdictional claims in published maps and institutional affiliations.

Printed on acid-free paper

This Springer imprint is published by the registered company Springer Nature Singapore Pte Ltd.
The registered company address is: 152 Beach Road, #21-01/04 Gateway East, Singapore 189721, Singapore

Acknowledgments

The authors are funded by the National Natural Science Foundation of China (31571080 and 31522025 to W. X.; 81771001 to Z. X.)

Contents

Chapter 1
A Brief Introduction

Wei Xiong

Abstract Auditory perception in mammals starts from the ear, the periphery part of the auditory system. The cochlear hair cells use their hair bundle to convert sound-induced mechanical vibration into receptor potential and then utilize ribbon synapse to convey the electrical signals to ascending spiral ganglion neurons. This biological process is called *auditory transduction* that is the first step and also the most essential part of hearing sensation. However, to achieve a competent processing capability of acoustic cues in large dynamics, high fidelity, and wide range, many astonishing molecular design and amazing biophysical assembly have been applied for the cochlear hair cells to fulfil the auditory transduction.

Keywords Auditory transduction · Mechanotransduction · Hair cells · Cochlea · Inner ear

1.1 Hair Cells Are Mechano-Electric and Can Be Electromotive

The discovery of hair-cell mechanotransduction (also called mechano-electrical transduction, MET) starts from study in invertebrates. Not like the case that a part of neuron senses the sensory input, such as free terminal of dorsal root ganglion neurons in the skin, researchers had believed there should be a unique type of receptor cells that are responsible for vibration sensation. Early in 1930s, auditory nerve fibre recording has shown that sound stimulation induced firing of spiral ganglion neurons [1]. The direct evidence that indicated the hair cells as the receptors came from the intracellular recording of hair cells. The hair cells directly gave synchronized receptor potential change in response to mechanical stimulation in the lateral

W. Xiong (✉)
School of Life Sciences, IDG/McGovern Institute for Brain Research, Tsinghua University, Beijing, China
e-mail: wei_xiong@mail.tsinghua.edu.cn

© The Author(s) 2018
W. Xiong, Z. Xu (eds.), *Mechanotransduction of the Hair Cell*,
SpringerBriefs in Biochemistry and Molecular Biology,
https://doi.org/10.1007/978-981-10-8557-4_1

line organ of the tail of the mudpuppy *Necturus maculosus* [2]. Later on, mechano-sensitivity of the cochlear hair cells was interrogated by intracellular recording in guinea pigs [3], a commonly used mammalian model. In isolated bullfrog saccule tissue, the properties of MET current were studied in details [4]. Nevertheless, whole-cell patch clamp was applied to hair cells to achieve a detailed information of the MET channel including some single-channel behaviour [5]. In general, no matter in lower vertebrates or higher mammals, the hair cells are the target cells in the organs that are responsible for either hearing or balance. Damage of hair cells immediately induces hearing or balance disorder. Undoubtedly, all the evidence indicates that the hair cell serves as the mechanosensor for special need of vibration-induced auditory or equilibrium perception.

In mammals, two types of auditory hair cells are found in cochlea. Inner hair cells (IHCs) are more similar to the hair cells in lower vertebrates. A new subtype of hair cells, outer hair cells (OHCs), has been developed to perform a new task – amplification – within cochlea [6]. By embedding a unique protein prestin in the plasma membrane, the OHCs gain robust electromotility on their cell bodies [7]. The length of OHC soma is longer when hyperpolarized and shorter when depolarized [8]. This function is super important for enhancing the shear stress towards hair bundle.

1.2   Cochlea Decodes Sound in Both Passive and Active Styles

To decompose the two major acoustic parameters, frequency and intensity, many strategies have been applied biologically. In lower vertebrates, an electrical tuning mechanism is mainly used [9]. It has been shown that calcium-activated potassium (BK) channels are expressed in the auditory hair cells with different isoforms along the basilar papilla [10]. Electrical oscillation of the hair cells has been tuned to certain frequencies depending on the different BK channel isoforms and densities of cell membrane [11]. However, in mammals, it is the basilar membrane that mainly decodes the frequencies physically. Only the hair cells on the vibrated part of basilar membrane can be activated. BK channel has been shown not contributing in hair-cell tuning in mammals [12].

Passive analysis of sound frequencies is indispensable for all hearing animals despite of realization forms. Other than that, mammals further develop an active process to enhance the mechanosensitivity [6]. Evolutionally, cochleae only exist in mammals and some non-mammalian vertebrates, e.g. birds. Mammalian OHCs possess motility at both cell body and hair bundle, which largely help the organ of Corti to gain stronger reactivity to enhance discrimination capability [13]. In general, the cochlear fluid and basilar membrane are the instruments as the passive frequency analyser, while OHCs provide active amplification to endow the inner ear with better amplification effect and frequency sensitivity.

1.3 Mammalian Inner Ear Is a Well-Evolved Bio-machine

Until now, the mammalian inner ear is recognized as a super biological machine in terms of its mechanic design. Firstly, it is a complicated bony labyrinth composed of cochlea, vestibule, and semi-circular cannels. Each part is specialized in structure to execute the desired function. To decode the acoustic cues, all frequency-selective hair cells are arrayed in the cochlea that is a long tubing but compactly coiled (2.5 turns for human cochlea). A general theory is that the coiled structure improves the length to allocate more sensory cells sitting on basilar membrane. Along this tubing, three compartments – scala tympani, scala media, and scala vestibuli – are divided by two cellular layers, basilar membrane and Reissner's membrane. Further, auditory hair cells are embedded in the organ of Corti inside scala media. Basilar membrane and tectorial membrane work cooperatively to help the hair cells to gain the best mechanosensitivity. Meanwhile the ribbon synapses decode big data of digitized sound information and transmit them to the central nerve system.

At the cellular level, dozens of proteins are recruited to maintain hair-cell functions, including MET. Currently, it has been characterized that more than 30 proteins play pivotal roles in hair cells. Nevertheless, there are at least 13 proteins (CDH23, PCDH15, USH1C, MYO7A, SANS, TMC1, TMC2, LHFPL5, TMIE, MYO15A, WHRN, CIB2, TOMT) known to be directly responsible for MET. These proteins will be introduced in the following chapters.

It is interesting that the transduction machinery invests a dense mass of resource to fulfil a relatively simple task – sampling mechanical signals. Anyhow, there must be necessary reason to do so. So far, it is not completely known yet what is the core pore-forming subunit and how the transducer complex works. In this Springer Brief, we would like to introduce most recent discoveries and embed them into our existing knowledge on MET of the hair cell.

References

1. Galambos, R., and H. Davis. 1943. The response of single auditory-nerve fibers to acoustic stimulation. *Journal of Neurophysiology* 6 (1): 39–57.
2. Harris, G.G., L.S. Frishkopf, and A. Flock. 1970. Receptor potentials from hair cells of the lateral line. *Science* 167 (3914): 76–79.
3. Russell, I.J., and P.M. Sellick. 1977. Tuning properties of cochlear hair cells. *Nature* 267 (5614): 858–860.
4. Hudspeth, A.J., and D.P. Corey. 1977. Sensitivity, polarity, and conductance change in response of vertebrate hair cells to controlled mechanical stimuli. *Proceedings of the National Academy of Sciences of the United States of America* 74 (6): 2407–2411.
5. Ohmori, H. 1985. Mechano-electrical transduction currents in isolated vestibular hair-cells of the chick. *The Journal of Physiology* 359 (Feb): 189–217.
6. Dallos, P. 1992. The active cochlea. *The Journal of Neuroscience* 12 (12): 4575–4585.
7. Zheng, J., et al. 2000. Prestin is the motor protein of cochlear outer hair cells. *Nature* 405 (6783): 149–155.

8. Ashmore, J.F. 1987. A fast motile response in guinea-pig outer hair cells: the cellular basis of the cochlear amplifier. *The Journal of Physiology* 388: 323–347.
9. Crawford, A.C., and R. Fettiplace. 1981. An electrical tuning mechanism in turtle cochlear hair cells. *The Journal of Physiology* 312: 377–412.
10. Ramanathan, K., et al. 1999. A molecular mechanism for electrical tuning of cochlear hair cells. *Science* 283 (5399): 215–217.
11. Fettiplace, R., and P.A. Fuchs. 1999. Mechanisms of hair cell tuning. *Annual Review of Physiology* 61: 809–834.
12. Oliver, D., et al. 2006. The role of BKCa channels in electrical signal encoding in the mammalian auditory periphery. *The Journal of Neuroscience* 26 (23): 6181–6189.
13. Hudspeth, A.J. 2014. Integrating the active process of hair cells with cochlear function. *Nature Reviews. Neuroscience* 15 (9): 600–614.

Chapter 2
Cellular Structure for Hair-Cell Mechanotransduction

Zhigang Xu

Abstract In this chapter, we will introduce the cellular structure that is responsible for hair-cell mechanotransduction (MET). Auditory hair cells reside in the inner ear, where they are interlaced in a precise pattern with various supporting cells. Hair cells got their name because each one has hundreds of hairy-looking membrane protrusions, namely, stereocilia, extending from its apical surface. Within each hair cell, various types of extracellular links couple different stereocilia with one another. Tip links connect the tips of shorter stereocilia to the lateral shaft of its taller neighbouring stereocilia, and the yet-unidentified MET channels localize at the tips of shorter stereocilia, near the lower end of tip links. When the stereocilia are deflected positively, tension of tip links increases, which changes the conformation of MET channels and increases their open probability. Driven by the potential difference and ion concentration difference, cations flux into hair cells through the MET channels and cause membrane depolarization.

Keywords Inner ear · Organ of Corti · Hair cells · Stereocilia · Tip links

2.1 Inner Ear and Hair Cells

The auditory system consists of three parts: the outer ear, the middle ear, and the inner ear (Fig. 2.1a). The outer ear funnels sound into the middle ear, which converts the airborne vibration into mechanical vibration and then transfers it to the cochlea of the inner ear. Besides the cochlea, in the inner ear, there is also the labyrinth, which belongs to the vestibular system (Fig. 2.1a). Here we are only concerned with the auditory hair cells in the cochlea, but we want to point out that the vestibular hair cells in the labyrinth have similar structure as auditory hair cells and

Z. Xu (✉)
School of Life Sciences, Shandong University, Jinan, Shandong, China
e-mail: xuzg@sdu.edu.cn

© The Author(s) 2018
W. Xiong, Z. Xu (eds.), *Mechanotransduction of the Hair Cell*,
SpringerBriefs in Biochemistry and Molecular Biology,
https://doi.org/10.1007/978-981-10-8557-4_2

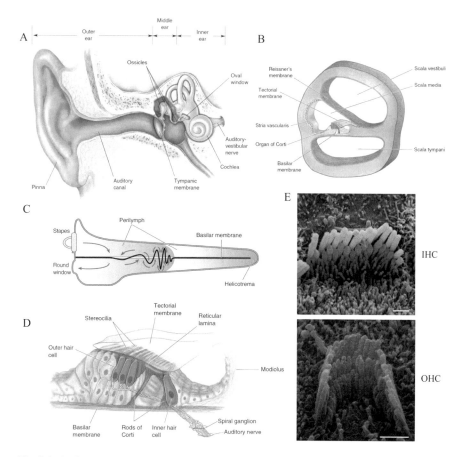

Fig. 2.1 Auditory pathway: the ear, organ of Corti, and hair cells. (**a**) The outer, middle, and inner ear. (**b**) The cross-section view of the cochlea. The cochlea contains three chambers: scala vestibuli, scala media, and scala tympani, which are separated by Reissner's membrane and the basilar membrane. The organ of Corti resides upon the basilar membrane. (**c**) A traveling wave in the basilar membrane. For the purpose of simplicity, the scala media and Reissner's membrane are not illustrated here. (**d**) The cross-section view of the organ of Corti. The organ of Corti sits upon the basilar membrane and is covered by the tectorial membrane. (**e**) Top view of the hair bundles of P8 mouse IHC and OHC. Scale bars, 1 μm
(**a**) through (**d**): Reprinted with permission from Bear MF. et al., Neuroscience: Exploring the brain (4th edition, 2015). Publisher: Wolters Kluwer. (**e**) Reprinted with permission from Wang Y. et al., *Front. Mol. Neurosci.*, 2017, 10: 401

that the molecular MET mechanism in auditory hair cells and vestibular hair cells is believed to be similar, if not the same.

The cochlea has a spiral shape, like a snail's shell. The inside of the cochlea is divided into three fluid-filled chambers: scala vestibuli, scala media, and scala tympani. Scala vestibuli is separated from scale media by Reissner's membrane, and scala media is separated from scala tympani by basilar membrane (Fig. 2.1b). Scala vestibuli and scala tympani are connected to each other at the apex of the cochlea through a hole called helicotrema, and these two chambers are filled with perilymph, which contains relatively low K^+ (7 mm) and high Na^+ (140 mm), similar to regular cerebrospinal fluid. In contrast, scala media is filled with endolymph, which contains high K^+ (150 mm) and low Na^+ (1 mm), similar to intracellular fluid. The ionic concentration differences and the selective permeability of Reissner's membrane result in the so-called endocochlear potential (EP): the electrical potential of endolymph is about 80 mV more positive than that of the perilymph.

The mechanical vibration transmitted from the middle ear moves the membrane that covers the oval window, an opening at the base of scala vestibuli. This movement displaces the perilymph and endolymph inside the cochlea and eventually causes the basilar membrane to move up and down. The basilar membrane starts its movement by bending near its base, which then propagates towards the apex as a "travelling wave" (Fig. 2.1c). Sitting on the basilar membrane is the organ of Corti, which harbours the auditory hair cells as well as various supporting cells (Fig. 2.1d). From base to apex along the cochlear duct, hair cells at different positions are tuned to different frequencies, which are known as characteristic frequencies. Hair cells at the base respond to high frequencies, and those at the apex respond to low frequencies [1, 2].

In mammals, there are two types of auditory hair cells: inner hair cells (IHCs) and outer hair cells (OHCs). Along the spiral cochlea, there are one single row of IHCs (about 3500 in human) and three rows of OHCs (about 12,000 in human). Hair cells got their name because each one has hundreds of hairy-looking membrane protrusions, namely, stereocilia, extending from its apical surface. IHCs have pear-shaped cell body and contain flattened U-shaped stereocilia. OHCs have cylinder-shaped cell body and contain V-shaped stereocilia (Fig. 2.1e). The tips of the OHC stereocilia are embedded in the overlying tectorial membrane.

IHCs form synapses with the spiral ganglion cells, which are afferent neurons. Spiral ganglion cells relay the auditory information to the cochlear nuclei in the medulla and eventually to the auditory cortex. OHCs, on the other hand, usually do not form synapses with the spiral ganglion. Instead, they mainly form synapses with efferent neurons and work as cochlear amplifier and play an important role in frequency tuning [3, 4]. Nevertheless, the molecular MET mechanisms in these two types of auditory hair cells are considered to be the same.

2.2 Hair Bundle: Stereocilia and Kinocilia

Hair cells are characteristic of their hairy-looking, F-actin-based stereocilia on the apical surface (Fig. 2.2a). One hair cell has dozens to hundreds of stereocilia, which are organized into several rows of increasing heights, forming a staircase-like pattern. The actin core in each stereocilium is tightly packed in a paracrystalline array, with the barbed (plus) ends pointing towards the stereociliary tips [5]. The actin core is dynamic in developing stereocilia but stable in mature stereocilia [6, 7]. The

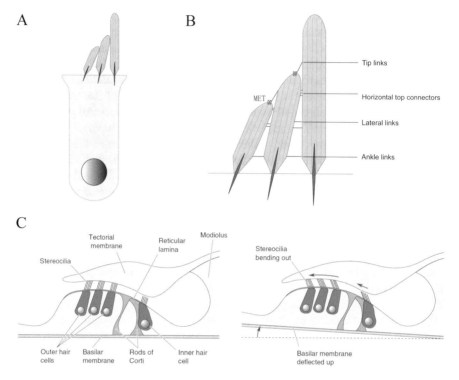

Fig. 2.2 Hair bundles: structure and function. (**a**) Schematic drawing of a hair cell. (**b**) Schematic drawing of extracellular links that connect stereocilia. (**c**) The model of sound-induced hair bundle deflection (Reprinted with permission from Bear MF. et al., Neuroscience: Exploring the brain (4th edition, 2015). Publisher: Wolters Kluwer)

stereocilium tapers at its base, and a few actin filaments (referred as rootlet) continue to insert into the F-actin matrix of the cuticular plate underneath the apical surface. This makes stereocilia easy to bend at the base when basilar membrane moves up and down (see below). Stereocilia are organized in a flattened U-shaped (IHCs) or V-shaped (OHCs) pattern, with the tallest stereocilia positioning at the vertices. The vertices of all stereocilia point away from the centre of the cochlea, establishing the planar cell polarity (PCP) in the cochlear epithelia.

Besides stereocilia, there is a single microtubule-based kinocilium on each hair cell's apical surface. Stereocilia and kinocilium together constitute the so-called hair bundle. Kinocilium localizes at the vertex of stereocilia, juxtaposed next to the tallest stereocilia. Kinocilium is degenerated at late developmental stage in cochlear hair cells (but persists in vestibular hair cells), suggesting that it is not necessary for MET [8–10]. Nevertheless, kinocilium is believed to play pivotal roles in stereocilia development and cochlear PCP establishment.

There are various types of extracellular links that couple different stereocilia with one another as well as with kinocilium [11, 12]. Four types of links connect stereocilia, namely, tip links, horizontal top connectors, lateral links, and ankle links. Among them, tip links and horizontal top connectors are found in the mature cochlear stereocilia, whereas lateral links and ankle links only temporally exist in developing cochlear stereocilia (Fig. 2.2b). The tallest stereocilia are connected to kinocilium by kinociliary links. As kinocilium only exist in developing auditory hair cells, kinociliary links in auditory hair cells are also transient. These links play important roles in the development and function of hair bundle.

The morphogenesis of hair bundle was first thoroughly studied in avian hair cells [5, 13]. At the onset of hair bundle morphogenesis, the apical surface of hair cell is covered uniformly with multiple microvilli, and a single kinocilium is localized in the centre of the apical surface. The microvilli elongate and form stereocilia of similar height then stop elongation and start to increase the width by adding more actin filaments. Meanwhile, the kinocilium moves from the centre to the bundle periphery at a random direction then reorients so that the kinocilia of all the hair cells are on approximately the same side of the cell. The stereocilia next to the kinocilium then start to elongate again, followed by elongation of the adjacent rows of stereocilia, forming a staircase-like pattern. After that, stereocilia stop elongation and increase their width. Kinocilium is lost during this period, and the unelongated stereocilia are eventually resorbed. The filaments in the central core of stereocilia extend basally to form rootlets. At last, stereocilia reinitiate elongation and grow to their final lengths. Development of mammalian auditory hair bundles generally follows the same theme, although the stages are less distinct [14, 15].

When sound-induced mechanical energy is transferred to the cochlea, the basilar membrane moves up and down. The tips of the OHC stereocilia are embedded in the tectorial membrane; hence, the movement of the basilar membrane relative to the tectorial membrane causes the deflection of the stereocilia of OHCs. Although the tips of IHC stereocilia are not embedded in the tectorial membrane, IHC stereocilia are also deflected, possibly by the moving endolymph. The extracellular links hold the stereocilia together, so all the stereocilia move as a unit (Fig. 2.2c). The movement of hair bundle provides the basis for hair-cell MET. The response of hair cells to stereocilia movement is direction-sensitive: movement towards the tallest stereocilia produces depolarized receptor potential, whereas opposite movement produces hyperpolarized receptor potential. There is a cosine relationship for responses to stereocilia movement of any directions, and the sensitive direction is referred as "axis of mechanical sensitivity" [16, 17].

2.3 Tip Links and Mechanotransduction

Among all the extracellular links that connect the stereocilia, tip links probably attract the most attention. Tip links connect the tips of shorter stereocilia to the lateral shaft of its taller neighbouring stereocilia [18]. Ultrastructural studies showed that they are 150–200 nm long and 8–11 nm thick, adopting a double-helical conformation with a 20–25 nm periodicity [19, 20] (Fig. 2.3a–c). Each filament in the double helix is composed of multiple globular structures that are 4 nm in diameter.

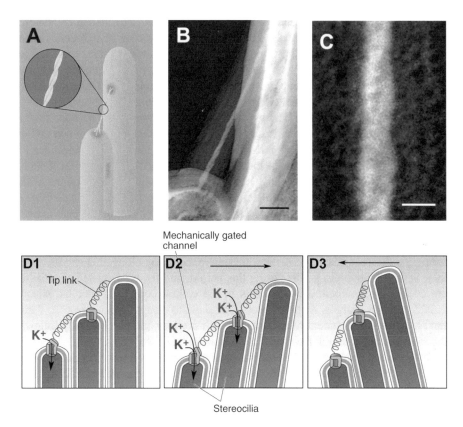

Fig. 2.3 Tip links: the pivotal linkers that gate the MET channels. (**a**) Model of tip-link structure. (**b**) Freeze-etch image of tip link from guinea pig hair cells. (**c**) Higher magnification view of the tip link in (**b**). (**d**) Model of MET mechanism (Scale bars: 50 nm in **b**; 10 nm in **c**)
(**a**) through (**c**): Reprinted with permission from Kachar B, et al. *Proc. Natl. Acad. Sci. U. S. A.*, 2000, 97(24):13336–41. Copyright (2000) National Academy of Sciences, U.S.A. (**d**) Reprinted with permission from Bear MF. et al., Neuroscience: Exploring the brain (4th edition, 2015). Publisher: Wolters Kluwer

Tip links are separated into two or three strands at both ends, which insert into the plasma membrane of stereocilia. Electron microscopy revealed that the upper and lower tip-link insertion sites display electron-dense plaques [21], which were named later as upper tip-link density (UTLD) and lower tip-link density (LTLD). Ca^{2+} chelators such as BAPTA cause disruption of tip links, suggesting that the structural integrity of tip links is Ca^{2+}-dependent [22]. Tip links were also shown to be asymmetric: the upper part and lower part are formed by different proteins [23], and the MET channels are placed near the lower end of tip links [24].

Several lines of evidence indicate that tip links are directly involved in the MET process. First, the direction of tip links is parallel to the bundle's axis of mechanical sensitivity [18]. Second, high-speed calcium imaging revealed that the MET channels are localized at the tips of shorter stereocilia, near the lower end of tip links [24].

Third, when tip links are disrupted with calcium chelators, MET is also abolished [22]. Fourth, hair-cell MET is interrupted in mutant mice that lose tip links [25–27].

Figure 2.3d summarizes our present understanding how MET happens in hair cells. When there is no sound stimuli, the stereocilia are not deflected and stand straight up. The resting tension of the tip links causes the MET channels to spend part of the time in the open state. Driven by the EP and the ionic concentration differences, a small amount of cations (K^+ and Ca^{2+}) enter hair cells through the channels (Fig. 2.3d1). When the stereocilia are deflected towards the taller edge, tension of the tip links increases, which in turn increases the open probability of the channels, and allows more cations to enter hair cells. The flux of cations causes membrane depolarization in hair cells (Fig. 2.3d2). On the other hand, when the stereocilia are deflected towards the shorter edge, tension of tip links decreases, less cations enter hair cells, and cell membrane is hyperpolarized (Fig. 2.3d3).

2.4 Discussion

Cochlear hair cells are the auditory receptor cells, and their characteristic stereocilia are indispensable for MET. In the present model, tip links, which connect the tips of shorter stereocilia to the lateral shaft of its taller neighbouring stereocilia, are placed in a central position next to the MET channels. Deflection of stereocilia changes tension of tip links, opens the MET channels, and produces receptor potential.

Given the importance of stereocilia in hair-cell MET, it is not surprising that deficits in stereocilia development or maintenance could lead to profound hearing loss. Indeed, mutations that affect stereociliary F-actin polymerization, bundling, or even actin itself have been shown to associate with syndromic or nonsyndromic deafness [28]. Kinocilia are not present in mature mammalian cochlear hair cells and hence are not necessary for MET. Nevertheless, kinocilia are essential for hair bundle development, as mutations that affect kinocilia are associated with hair bundle polarity deficits [29, 30].

The small numbers of hair cells and the scarcity of functional MET channels in each hair cell hindered the identification of the channel. We now know that the MET channels localize at the tips of shorter stereocilia, close to the lower insertion site of tip links. The property and molecular composition of the MET machinery are discussed in the following chapters.

References

1. Liberman, M.C. 1982. The cochlear frequency map for the cat: labeling auditory-nerve fibers of known characteristic frequency. *The Journal of the Acoustical Society of America* 72 (5): 1441–1449.

2. Muller, M. 1991. Frequency representation in the rat cochlea. *Hearing Research* 51 (2): 247–254.

3. Kiang, N.Y., et al. 1986. Single unit clues to cochlear mechanisms. *Hearing Research* 22: 171–182.

4. Dallos, P. 1992. The active cochlea. *The Journal of Neuroscience* 12 (12): 4575–4585.

5. Tilney, L.G., M.S. Tilney, and D.J. DeRosier. 1992. Actin filaments, stereocilia, and hair cells: how cells count and measure. *Annual Review of Cell Biology* 8: 257–274.

6. Zhang, D.S., et al. 2012. Multi-isotope imaging mass spectrometry reveals slow protein turnover in hair-cell stereocilia. *Nature* 481 (7382): 520–524.

7. Drummond, M.C., et al. 2015. Live-cell imaging of actin dynamics reveals mechanisms of stereocilia length regulation in the inner ear. *Nature Communications* 6: 6873.

8. Lindeman, H.H., et al. 1971. The sensory hairs and the tectorial membrane in the development of the cat s organ of Corti. A scanning electron microscopic study. *Acta Oto-Laryngologica* 72 (4): 229–242.

9. Tanaka, K., and C.A. Smith. 1978. Structure of the chicken's inner ear: SEM and TEM study. *The American Journal of Anatomy* 153 (2): 251–271.

10. Denman-Johnson, K., and A. Forge. 1999. Establishment of hair bundle polarity and orientation in the developing vestibular system of the mouse. *Journal of Neurocytology* 28 (10–11): 821–835.

11. Goodyear, R., and G. Richardson. 1992. Distribution of the 275 kD hair cell antigen and cell surface specialisations on auditory and vestibular hair bundles in the chicken inner ear. *The Journal of Comparative Neurology* 325 (2): 243–256.

12. Goodyear, R.J., et al. 2005. Development and properties of stereociliary link types in hair cells of the mouse cochlea. *The Journal of Comparative Neurology* 485 (1): 75–85.

13. Tilney, L.G., and M.S. Tilney. 1986. Functional organization of the cytoskeleton. *Hearing Research* 22: 55–77.

14. Kaltenbach, J.A., P.R. Falzarano, and T.H. Simpson. 1994. Postnatal development of the hamster cochlea. II. Growth and differentiation of stereocilia bundles. *the Journal of Comparative Neurology* 350 (2): 187–198.

15. Zine, A., and R. Romand. 1996. Development of the auditory receptors of the rat: a SEM study. *Brain Research* 721 (1–2): 49–58.

16. Flock, A. 1964. Electron microscopic and electrophysiological studies on the lateral line canal organ. *Acta Oto-Laryngologica. Supplementum* SUPPL 199: 1–90.

17. Shotwell, S.L., R. Jacobs, and A.J. Hudspeth. 1981. Directional sensitivity of individual vertebrate hair cells to controlled deflection of their hair bundles. *Annals of the New York Academy of Sciences* 374: 1–10.

18. Pickles, J.O., S.D. Comis, and M.P. Osborne. 1984. Cross-links between stereocilia in the guinea pig organ of Corti, and their possible relation to sensory transduction. *Hearing Research* 15 (2): 103–112.

19. Kachar, B., et al. 2000. High-resolution structure of hair-cell tip links. *Proceedings of the National Academy of Sciences of the United States of America* 97 (24): 13336–13341.

20. Tsuprun, V., R.J. Goodyear, and G.P. Richardson. 2004. The structure of tip links and kinocilial links in avian sensory hair bundles. *Biophysical Journal* 87 (6): 4106–4112.

21. Furness, D.N., and C.M. Hackney. 1985. Cross-links between stereocilia in the guinea pig cochlea. *Hearing Research* 18 (2): 177–188.

22. Assad, J.A., G.M. Shepherd, and D.P. Corey. 1991. Tip-link integrity and mechanical transduction in vertebrate hair cells. *Neuron* 7 (6): 985–994.

23. Kazmierczak, P., et al. 2007. Cadherin 23 and protocadherin 15 interact to form tip-link filaments in sensory hair cells. *Nature* 449 (7158): 87–91.

24. Beurg, M., et al. 2009. Localization of inner hair cell mechanotransducer channels using high-speed calcium imaging. *Nature Neuroscience* 12 (5): 553–558.

25. Schwander, M., et al. 2009. A mouse model for nonsyndromic deafness (DFNB12) links hearing loss to defects in tip links of mechanosensory hair cells. *Proceedings of the National Academy of Sciences of the United States of America* 106 (13): 5252–5257.
26. Alagramam, K.N., et al. 2011. Mutations in protocadherin 15 and cadherin 23 affect tip links and mechanotransduction in mammalian sensory hair cells. *PLoS One* 6 (4): e19183.
27. Geng, R., et al. 2013. Noddy, a mouse harboring a missense mutation in protocadherin-15, reveals the impact of disrupting a critical interaction site between tip-link cadherins in inner ear hair cells. *The Journal of Neuroscience* 33 (10): 4395–4404.
28. Barr-Gillespie, P.G. 2015. Assembly of hair bundles, an amazing problem for cell biology. *Molecular Biology of the Cell* 26 (15): 2727–2732.
29. Ross, A.J., et al. 2005. Disruption of Bardet-Biedl syndrome ciliary proteins perturbs planar cell polarity in vertebrates. *Nature Genetics* 37 (10): 1135–1140.
30. Jones, C., et al. 2008. Ciliary proteins link basal body polarization to planar cell polarity regulation. *Nature Genetics* 40 (1): 69–77.

Chapter 3
Biophysical Properties
of Mechanotransduction

Wei Xiong

Abstract In early time, the coarse signals, such as cochlear microphonics and sum-mating potentials, were recorded by electrocochleography by placing a metal elec-trode on the round window, which reflected the sound-induced electrical responses mostly mediated by hair cells. However, these signals are mainly a summated response from a group of hair cells. The direct evidence came from the intracellular recording of hair cells in the tail lateral line of mudpuppy *Necturus maculosus* (Harris et al., Science 167(3914):76–79, 1970). Similarly, auditory response from the cochlear hair cells was probed by intracellular recording in guinea pig (Russell and Sellick, Nature 267(5614):858–860, 1977; J Physiol 284:261–290, 1978). In isolated bullfrog saccule tissue, the mechanotransduction (MET) current was recorded in hair cells, which provided the first evidence that the deflection of hair bundle induced receptor potential change of hair cells (Hudspeth and Corey, Proc Natl Acad Sci USA 74(6):2407–2411, 1977). Nevertheless, whole-cell patch clamp was applied to hair cells to achieve a detailed information of the MET channel including some single-channel behaviour (Ohmori, J Physiol 359:189–217, 1985). From then on, researchers have studied most of the biophysical properties of the channel systematically by electrophysiology, pharmacology, and optical imaging without knowing the molecular identity of the MET channel. Now several important questions have been tackled in this chapter, including the following: Where does the channels localize in the hair cells? What kind of ions do the channels pass through? How are the channels activated and then adapted? How many channels are there opened per tip link? What are the single-channel properties?

Keywords Channel · Permeability · Activation · Adaptation · Reverse polarity

W. Xiong (✉)
School of Life Sciences, IDG/McGovern Institute for Brain Research, Tsinghua University,
Beijing, China
e-mail: wei_xiong@mail.tsinghua.edu.cn

© The Author(s) 2018   15
W. Xiong, Z. Xu (eds.), *Mechanotransduction of the Hair Cell*,
SpringerBriefs in Biochemistry and Molecular Biology,
https://doi.org/10.1007/978-981-10-8557-4_3

3.1 Channel Localization

It is not surprising to imagine that the cilia-based hair bundle is developed in the hair cells for the vibration sensation. In addition, the filamentous tip-link structure was observed between a stereocilium and its taller neighbouring stereocilium, which was proposed to play roles in transduction. An intriguing hypothesis was that the hair bundle hosts the MET channels and allocates the channels specifically to be around the tip links. However, the problem was how to prove it experimentally. Three decades ago, several pioneer research teams have used very delicate methods to locate the MET channels on hair bundle in a subcellular scale. In 1982, Hudspeth applied extracellular recording to draw a heat map for the channel distribution in the hair bundle of bullfrog saccular hair cells. Operationally, a fine-tip electrode was placed at an array of sites on the hair bundle to probe the flow of transduction current. By this way, the maximal transducer response took place at or near the top of the hair bundle, i.e. the distal ends of the stereocilia [6]. Late on, Jaramillo and Hudspeth used another strategy to further confirm the localization of the transducer channels at the hair bundles. This time, they locally applied a channel blocker gentamicin to scan the possible hot spot that can inhibit the transduction currents. Similarly, the most sensitive site for channel blocker was at the top of the hair bundle [7]. These studies suggested that the top surface of hair bundles was the most sensitive part to mechanical stimulation and aminoglycoside inhibition. However, it still remained puzzled whether the MET channels were exactly here for the functionality. To really point out where the ion fluxes in, cellular calcium imaging was recruited to visualize the channel activity; however, the epifluorescence microscopy just gave an ambiguous result due to low resolution [8]. Late on the confocal microscopy, a state-of-the-art technology back at that time was used to address this question. The higher temporal and spatial resolution endowed the confocal microscopy with the power to solve the question. The focal scan and line scan both showed that the calcium enters from the very top of the hair bundle and then diffuses along the stereocilia [9, 10], which is exactly the location of tip links.

Then the emerging question is how does the channels distribute around the tip link, symmetrically at both ends of the tip link or asymmetrically at one end of the tip link? If it is one end, then which end should host the channel, i.e. upper tip-link density (UTLD) or lower tip-link density (LTLD)? In previous studies, chicken and frog hair cells were used as model. The structure of lower vertebrate hair cells was similar to mammalian vestibular hair cells that had less manifestation of polarity. The pattern of calcium influx was not clear enough to draw a crystal conclusion that the MET channels localized on either or both end of the tip links [9]. It was until 2009 that the accurate location was firmly determined by Fettiplace and Ricci groups. They used rats as a model since the mammalian cochlear hair cells possessed three rows of stereocilia that showed an obvious staircase structure. Utilizing an ultrafast swept-field confocal microscope and calcium indicator with fast kinetics, they showed that the calcium influx only happened from the top of the second and third rows of stereocilia while leaving the tallest one intact during mechanical deflection. Then it was assumed that the channel is at the lower tip-link side both in IHCs and OHCs [11]. Therefore, the asymmetry applied to not only the bundle structure but also the molecular distribution including the MET channels.

3.2   Channel Selectivity and Permeability

A follow-up question was which ions passed through the transducer channels in hair cells? Early in 1979, Corey and Hudspeth used adult bullfrog preparation to probe this question. Back at that time, dual sharp electrodes were applied with a simple voltage clamp configuration. The hair bundle was deflected with a triangle pattern at a magnitude of 1–2 μm and frequency of 10 Hz. By exchanging ionic solution on the apical side but keeping the basolateral side in perilymph bath, the selectivity of the transducer channel was examined as a nonselective cation channel. Comparing to K^+, the relative permeability reflected by microphonic current was 0.9 by Li^+, 0.9 by Na^+, 1.0 by Rb^+, and 1.0 by Cs^+. And more interestingly, ammonium ion (NH_4^+) was 1.3 [12]. Calcium seemed to be an important cofactor of MET current with Sr^{2+} as a replacement to Ca^{2+}, but Mg^{2+} and Ba^{2+} do not [5, 12].

Aminoglycoside has been a well-known toxic to hair cells. A major effect pathway was that aminoglycoside entered the hair cell through the MET channels and blocked the channels [5]. Despite of the ototoxicity of aminoglycoside, these molecules also provided an insight how big the channel was. It means that the channel pore is big enough to let a molecule such as streptomycin in. Interestingly, FM1-43 that has been intensively used in monitoring vesicle trafficking was shown to block the channels and compete with aminoglycoside [13]. As a summary, Farris et al. made a series of pharmacological examination of MET channel with multiple antagonists known targeting those common channels. The pharmacological profile indicated that the MET channel was closer to cyclic nucleotide-gated (CNG) and transient receptor potential (TRP) channels [14].

Further, many components were verified to change the channel transduction. Depletion of PIP2 in hair cells reduced the channel amplitude [15, 16]. Also it was reported that loss of TMC proteins caused an altered calcium permeability in hair cells [17]. Corey and Hudspeth checked the permeability of an organic cation TMA that had a size of 0.54 nm in diameter [12]. This evidence indicated that the transducer channels had an internal diameter of at least 0.65 nm. Farris et al. systematically discussed the appropriate pore size by testing a series of small organic molecules, with an idea that the narrowest diameter of the pore was 1.25 nm. The channel was around 3.1 nm in length and less than 1.7 nm in width [14].

3.3 Activation and Adaptation

Adaptation is broadly used as a general concept for sensory cue processing that the organisms gain a distinguished capability to collect useful information out of noise. In terms of each hair cell, it strongly adapts to sustained mechanical stimuli in millisecond time constant. The first piece of work on hair-cell adaptation has been studied systematically in bullfrog sacculus hair cells [18]. By directly exposing hair cells from frog, the nerve activity and MET current were examined by in vivo recording. The discharge rate of saccular nerve was increased during onset and termination of acceleration stimulation. The displacement-response curve of MET was not changed obviously in shape but only shifted when superimposing a step-like stimulation. This evidence strongly showed that an adaptation existed in hair cells together with many studies from others [19–24]. And this millisecond-level adaptation was defined as slow adaptation since the channel was later recognized to have amazingly fast kinetics.

It was possible to be studied more deeply when introducing a novel type of piezoelectric actuator with microsecond responsivity [24–27]. Surrendered to a step-like deflection of hair bundle, there were two components of adaptation after the activation phase [27]. The manifestation of currents showed dramatic kinetics difference with previously observed current property. Immediately following a rapid activation (usually less than 100 μs), there is a fast adaptation that is quite similar to fast inactivation of voltage-gated sodium channel but with a time constant of sub-millisecond in mammalian hair cells. Then there is a slow adaptation [25, 28]. It is not sure whether this fast activation time was still underestimated due to physical limitation of stimulation apparatus. The channel opens and recloses so rapidly that it is consistent with the coding requirement of the sound information, especially for middle-to-high frequencies. In physiological status, sinusoid sound wave seems not challenging the MET channels too much. However, it is still an interesting question how high the frequency hair cell can detect by itself, such as whether a 10 kHz hair cell responses accurately to the 10 kHz wave. It means that the rising/falling phase of stimulation is around 25 μs theoretically, though the basilar membrane has done most of the job to analyse the frequencies.

Adaptation can be affected by many factors. Early study has found the adaptation was sensitive to voltage and calcium [19, 20, 29]. It was mainly resulted from calcium inhibition to the channel [30]. Recently, many MET components were characterized as essential regulator to the channel kinetics. It has been reported that Myo7a, harmonin, and LHFPL5 (also known as TMHS) ablation reduced the fast adaptation [31–33]. Actually membrane lipid around is also the important player for normal function of MET channels. A representative case is PIP2 that deeply contributed to current kinetics, such as conductance and adaptation [15, 16].

3.4 Single-Channel Properties

The channels are highly clustered at LTLD, so the number of tip links determines the number of active channels. By manipulating the number of the tip links to a few, it provided an opportunity to study the MET channel at single-channel level. A step-like MET current has been observed with a triangular stimulation in whole-cell patch-clamped chicken hair cells [34]. It was considered as single-channel recording of transducer channels that was later studied intensively by several groups though the conductance was reported ranging from 10 pS to 110 pS [5, 34–39]. Especially, Crawford et al. proposed that the conductance is around 110 pS in turtle hair cells, and late Geleoc et al. had a similar observation on mice [36, 37]. In 2003, Ricci et al. systematically analysed the single-channel behaviour of MET complex in turtle. The single-channel events were perfectly matching the macroscopic current kinetics after average assembly. Extracellular calcium deprivation from 2.8 mm to 0.05 mm increased the channel conductance from 118 pS to 215 pS by average [38]. There was a tonotopic distribution of channel conductances both in turtle [38] and in rat [39]. However, IHCs show no tonotopic variation on MET channel conductances that is equal to maximal conductance of OHCs, known as high-frequency hair cells.

With a relatively accurate single-channel conductance measurement, it is easy to calculate number of the channels per tip link. In 2006, Beurg et al. measured number of stereocilia and amplitude of transducer current. By average, there was 91 stereocilia that represented 60 tip links per OHC in middle coil. By measuring the saturated MET current as 1.2 nA and the single channel current as 12.1 pA, it was counted 1.65 channel per tip link. It was also confirmed that there are two channels per tip link by Ca^{2+} imaging calculation. By validating number of active stereocilia in IHCs with calcium imaging, the linearized fit showed 35.4 pA MET current per stereocilium. Consider the single-channel conductance as 15 pA at −80 mV, there were estimated two channels per tip link [11, 39].

3.5 Reverse-Polarity Mechanotransduction

In mature hair cells, deflection of the hair bundles towards the tallest stereocilia increases the open probability of the sensory MET channels, while deflection in the opposite direction decreases the open probability [40]. However, in some conditions, there is a type of MET current other than the classic properties of transducer channels [41–47]. Controlled by a sinusoid wave, a fluid jet generated a MET current at the negative phase in addition to the positive phase [42, 46]. In TMC1 and TMC2 double knockout mice, the hair cells showed this type of MET current even there was no classic MET current once the hair bundle was negatively deflected intensively (hence short as reverse-polarity MET current) [41]. This reverse-polarity current came out also when the hair bundle was treated by BAPTA, a calcium chelator breaking up tip links [41, 47]. It seems that the reverse-polarity current showed up once the MET complex was dissembled or immature [48]. This reminds us a series of observation that MET current was recorded in the inhibitory direction in null mice with MET component deficit [41–47] and a developing hair cell showed a lack of directional sensitivity [47–49]. The ion selectivity and responsiveness to pharmacological blockers of the reverse-polarity current are similar but not identical to that of the regular MET current [41, 47, 50]. High-speed Ca^{2+} imaging suggested that the reverse-polarity channels are not localized to the hair bundle but distributed at the apical surface of hair cells [48].

Then a debate arose that whether this reverse-polarity channel is identical to the classic MET channel. A new type of MET channel might be responsible for this reverse-polarity current, or the two channels are the same one but just locating at different position to give different biophysical behaviour. The TMC1 and TMC2 double knockout hair cells present the reverse-polarity current, which suggests either the two types of MET currents were from two types of the channels or the TMCs are not the channels for reverse polarity. In 2016, Mueller lab excluded the possibility of one-channel hypothesis. They found Piezo2 is specifically expressed in OHCs, while Piezo1 is not detected in hair cells. It has been well known that Piezos are MET channels that play pivotal roles in proprioception, sheath stress, lung airway, and so on. The molecular mechanism of Piezo2 in hair-cell MET is discussed in Chap. 4. Since Piezo2 null mice were embryonic lethal, they made conditional knockout by introducing Pax2-specific deletion of Piezo2 in the inner ear cells but keep the animal alive. The Piezo2 conditional knockout mice showed classic MET currents and relatively normal auditory function but only lost the reverse-polarity current. It is an amazing phenotype though we still have no idea about what is the physiological function of reverse-polarity current. Ironically, we have known so many properties about auditory MET, but the molecular identity is still elusive. Of course, it have placed TMCs again the first candidate for the MET channel.

3.6 Discussion

In this chapter, we summarized the biophysical properties for the mechanotransducer channels of hair cells. As a nonselective cation channel, it is highly restricted to the membrane patch proximal to the lower end of the tip link in mature hair cells. It generally allows most of the regular cations to flux in, but calcium inhibits the channel from intracellular side once it comes in. With the modulation of calcium and scaffold proteins, the channels are endowed with kinetics of fast and slow adaptation. Surprisingly, a type of "reverse-polarity" MET current exists in addition to the classic MET current in hair cells. Clearly, they are resulted from different channel proteins, especially known as Piezo2 contributing on reverse-polarity MET. However, our biophysical understanding of the classic MET channel is still limited by the fact that the channel was not cloned yet, or at least the channel cannot be reconstructed in an exogenously expressing system. TMCs are the top candidates, but their role in MET complex is still in debate. To know everything about the MET of the hair cells, we need significant input from molecular genetics and biochemistry. Of course, it is very inspiring to get a whole picture by embedding abundant morphological and biophysical knowledge with molecular basis of MET machinery that will be discussed in next chapter.

References

1. Harris, G.G., L.S. Frishkopf, and A. Flock. 1970. Receptor potentials from hair cells of the lateral line. *Science* 167 (3914): 76–79.
2. Russell, I.J., and P.M. Sellick. 1977. Tuning properties of cochlear hair cells. *Nature* 267 (5614): 858–860.
3. ———. 1978. Intracellular studies of hair cells in the mammalian cochlea. *The Journal of Physiology* 284: 261–290.
4. Hudspeth, A.J., and D.P. Corey. 1977. Sensitivity, polarity, and conductance change in the response of vertebrate hair cells to controlled mechanical stimuli. *Proceedings of the National Academy of Sciences of the United States of America* 74 (6): 2407–2411.
5. Ohmori, H. 1985. Mechano-electrical transduction currents in isolated vestibular hair cells of the chick. *The Journal of Physiology* 359: 189–217.
6. Hudspeth, A.J. 1982. Extracellular current flow and the site of transduction by vertebrate hair cells. *The Journal of Neuroscience* 2 (1): 1–10.
7. Jaramillo, F., and A.J. Hudspeth. 1991. Localization of the hair-cells transduction channels at the hair bundles top by iontophoretic application of a channel blocker. *Neuron* 7 (3): 409–420.
8. Ohmori, H. 1988. Mechanical stimulation and Fura-2 fluorescence in the hair bundle of dissociated hair cells of the chick. *The Journal of Physiology* 399: 115–137.
9. Denk, W., et al. 1995. Calcium imaging of single stereocilia in hair cells: localization of transduction channels at both ends of tip links. *Neuron* 15 (6): 1311–1321.
10. Lumpkin, E.A., and A.J. Hudspeth. 1995. Detection of Ca2+ entry through mechanosensitive channels localizes the site of mechanoelectrical transduction in hair cells. *Proceedings of the National Academy of Sciences of the United States of America* 92 (22): 10297–10301.
11. Beurg, M., et al. 2009. Localization of inner hair cell mechanotransducer channels using high-speed calcium imaging. *Nature Neuroscience* 12 (5): 553–558.
12. Corey, D.P., and A.J. Hudspeth. 1979. Ionic basis of the receptor potential in a vertebrate hair cell. *Nature* 281 (5733): 675–677.
13. Gale, J.E., et al. 2001. FM1-43 dye behaves as a permeant blocker of the hair-cell mechanotransducer channel. *Journal of Neuroscience* 21 (18): 7013–7025.
14. Farris, H.E., et al. 2004. Probing the pore of the auditory hair cell mechanotransducer channel in turtle. *The Journal of Physiology* 558 (Pt 3): 769–792.
15. Hirono, M., et al. 2004. Hair cells require phosphatidylinositol 4,5-bisphosphate for mechanical transduction and adaptation. *Neuron* 44 (2): 309–320.
16. Effertz, T., et al. 2017. Phosphoinositol-4,5-bisphosphate regulates auditory hair-cell mechanotransduction-channel pore properties and fast adaptation. *The Journal of Neuroscience* 37 (48): 11632–11646.

17. Kim, K.X., and R. Fettiplace. 2013. Developmental changes in the cochlear hair cell mechanotransducer channel and their regulation by transmembrane channel-like proteins. *The Journal of General Physiology* 141 (1): 141–148.
18. Eatock, R.A., D.P. Corey, and A.J. Hudspeth. 1987. Adaptation of mechanoelectrical transduction in hair-cells of the Bullfrogs Sacculus. *Journal of Neuroscience* 7 (9): 2821–2836.
19. Assad, J.A., N. Hacohen, and D.P. Corey. 1989. Voltage dependence of adaptation and active bundle movement in bullfrog saccular hair cells. *Proceedings of the National Academy of Sciences of the United States of America* 86 (8): 2918–2922.
20. Crawford, A.C., M.G. Evans, and R. Fettiplace. 1989. Activation and adaptation of transducer currents in turtle hair cells. *The Journal of Physiology* 419: 405–434.
21. Assad, J.A., and D.P. Corey. 1992. An active motor model for adaptation by vertebrate hair cells. *The Journal of Neuroscience* 12 (9): 3291–3309.
22. Shepherd, G.M., and D.P. Corey. 1994. The extent of adaptation in bullfrog saccular hair cells. *The Journal of Neuroscience* 14 (10): 6217–6229.
23. Holt, J.R., D.P. Corey, and R.A. Eatock. 1997. Mechanoelectrical transduction and adaptation in hair cells of the mouse utricle, a low-frequency vestibular organ. *The Journal of Neuroscience* 17 (22): 8739–8748.
24. Ricci, A.J., Y.C. Wu, and R. Fettiplace. 1998. The endogenous calcium buffer and the time course of transducer adaptation in auditory hair cells. *The Journal of Neuroscience* 18 (20): 8261–8277.
25. Kennedy, H.J., et al. 2003. Fast adaptation of mechanoelectrical transducer channels in mammalian cochlear hair cells. *Nature Neuroscience* 6 (8): 832–836.
26. Ricci, A.J., and R. Fettiplace. 1998. Calcium permeation of the turtle hair cell mechanotransducer channel and its relation to the composition of endolymph. *The Journal of Physiology* 506 (Pt 1): 159–173.
27. Wu, Y.C., A.J. Ricci, and R. Fettiplace. 1999. Two components of transducer adaptation in auditory hair cells. *Journal of Neurophysiology* 82 (5): 2171–2181.
28. Ricci, A.J., et al. 2005. The transduction channel filter in auditory hair cells. *The Journal of Neuroscience* 25 (34): 7831–7839.
29. Hacohen, N., et al. 1989. Regulation of tension on hair-cell transduction channels: displacement and calcium dependence. *The Journal of Neuroscience* 9 (11): 3988–3997.
30. Kimitsuki, T., and H. Ohmori. 1992. The effect of caged calcium release on the adaptation of the transduction current in chick hair cells. *The Journal of Physiology* 458: 27–40.
31. Xiong, W., et al. 2012. TMHS is an integral component of the mechanotransduction machinery of cochlear hair cells. *Cell* 151 (6): 1283–1295.
32. Grillet, N., et al. 2009. Harmonin mutations cause mechanotransduction defects in cochlear hair cells. *Neuron* 62 (3): 375–387.
33. Kros, C.J., et al. 2002. Reduced climbing and increased slipping adaptation in cochlear hair cells of mice with Myo7a mutations. *Nature Neuroscience* 5 (1): 41–47.
34. Ohmori, H. 1984. Mechanoelectrical transducer has discrete conductances in the chick vestibular hair cell. *Proceedings of the National Academy of Sciences of the United States of America* 81 (6): 1888–1891.
35. Holton, T., and A.J. Hudspeth. 1986. The transduction channel of hair cells from the bull-frog characterized by noise analysis. *The Journal of Physiology* 375: 195–227.
36. Crawford, A.C., M.G. Evans, and R. Fettiplace. 1991. The actions of calcium on the mechanoelectrical transducer current of turtle hair cells. *The Journal of Physiology* 434: 369–398.
37. Geleoc, G.S., et al. 1997. A quantitative comparison of mechanoelectrical transduction in vestibular and auditory hair cells of neonatal mice. *Proceedings of the Biological Sciences* 264 (1381): 611–621.
38. Ricci, A.J., A.C. Crawford, and R. Fettiplace. 2003. Tonotopic variation in the conductance of the hair cell mechanotransducer channel. *Neuron* 40 (5): 983–990.
39. Beurg, M., et al. 2006. A large-conductance calcium-selective mechanotransducer channel in mammalian cochlear hair cells. *The Journal of Neuroscience* 26 (43): 10992–11000.

40. Shotwell, S.L., R. Jacobs, and A.J. Hudspeth. 1981. Directional sensitivity of individual verte-brate hair cells to controlled deflection of their hair bundles. *Annals of the New York Academy of Sciences* 374: 1–10.
41. Kim, K.X., et al. 2013. The role of transmembrane channel-like proteins in the operation of hair cell mechanotransducer channels. *The Journal of General Physiology* 142 (5): 493–505.
42. Alagramam, K.N., et al. 2011. Mutations in protocadherin 15 and cadherin 23 affect tip links and mechanotransduction in mammalian sensory hair cells. *PLoS One* 6 (4): e19183.
43. Zhao, B., et al. 2014. TMIE is an essential component of the mechanotransduction machinery of cochlear hair cells. *Neuron* 84 (5): 954–967.
44. Beurg, M., et al. 2015. Subunit determination of the conductance of hair-cell mechanotrans-ducer channels. *Proceedings of the National Academy of Sciences of the United States of America* 112 (5): 1589–1594.
45. Michalski, N., et al. 2007. Molecular characterization of the ankle-link complex in cochlear hair cells and its role in the hair bundle functioning. *The Journal of Neuroscience* 27 (24): 6478–6488.
46. Stepanyan, R., and G.I. Frolenkov. 2009. Fast adaptation and Ca2+ sensitivity of the mecha-notransducer require myosin-XVa in inner but not outer cochlear hair cells. *The Journal of Neuroscience* 29 (13): 4023–4034.
47. Marcotti, W., et al. 2014. Transduction without tip links in cochlear hair cells is mediated by ion channels with permeation properties distinct from those of the mechano-electrical trans-ducer channel. *The Journal of Neuroscience* 34 (16): 5505–5514.
48. Beurg, M., et al. 2016. Development and localization of reverse-polarity mechanotransducer channels in cochlear hair cells. *Proceedings of the National Academy of Sciences of the United States of America* 113 (24): 6767–6772.
49. Waguespack, J., et al. 2007. Stepwise morphological and functional maturation of mechano-transduction in rat outer hair cells. *The Journal of Neuroscience* 27 (50): 13890–13902.
50. Beurg, M., K.X. Kim, and R. Fettiplace. 2014. Conductance and block of hair-cell mechano-transducer channels in transmembrane channel-like protein mutants. *The Journal of General Physiology* 144 (1): 55–69.

Chapter 4
Molecular Components
of Mechanotransduction Machinery

Zhigang Xu

Abstract After decades of intense investigation, the molecular components of mammalian hair-cell mechanotransduction (MET) machinery have started to emerge. Convincing evidences suggested that tip links are composed of two atypical cadherin proteins, protocadherin 15 (PCDH15) and cadherin 23 (CDH23). Meanwhile, the identity of the MET channel is still not confirmative, although several promising candidates have been put forward. In this chapter, we will first introduce the recent progress of our understanding of tip links, as well as the so-called upper and lower tip-link complexes associated with them. Then we will focus on the MET channel that lies at the heart of the MET machinery. TMC1, TMC2, LHFPL5, TMIE, and CIB2 have been suggested to be integral components of the machinery, but confirmative evidences for them as the pore-forming subunits of the channel are still missing. Lastly, we will briefly discuss the recent identification of PIEZO2 as the channel responsible for the reverse-polarity MET currents.

Keywords Hair cells · Tip links · MET machinery

4.1 Tip-Link Proteins

Tip links play a critical role in hair-cell MET. The molecular composition of tip links had been the subject of intense debate for decades, until two atypical cadherins, cadherin 23 (CDH23) and protocadherin 15 (PCDH15), were identified as tip-link components [1–3]. CDH23 and PCDH15 bind to each other via their extracellular N-terminal ends and make up the upper part and the lower part of tip links, respectively. Their cytoplasmic ends interact with other stereociliary proteins and form the so-called upper tip-link complex and lower tip-link complex.

Z. Xu (✉)
School of Life Sciences, Shandong University, Jinan, Shandong, China
e-mail: xuzg@sdu.edu.cn

© The Author(s) 2018
W. Xiong, Z. Xu (eds.), *Mechanotransduction of the Hair Cell*,
SpringerBriefs in Biochemistry and Molecular Biology,
https://doi.org/10.1007/978-981-10-8557-4_4

25

Table 4.1 Known USH genes and proteins

USH type	Locus	Gene	Protein	Description	References
USH1	USH1B	*MYO7A*	Myosin VIIA	Unconventional myosin	[10]
	USH1C	*USH1C*	Harmonin	PDZ scaffold protein	[11, 12]
	USH1D	*CDH23*	Cadherin 23	Atypical cadherin	[4, 5]
	USH1F	*PCDH15*	Protocadherin 15	Atypical cadherin	[6, 7]
	USH1G	*USH1G*	SANS	Scaffold protein	[13]
USH2	USH2A	*USH2A*	Usherin	Transmembrane protein	[14]
	USH2C	*ADGRV1*	ADGRV1	G-protein-coupled receptor	[15]
	USH2D	*WHRN*	Whirlin	PDZ scaffold protein	[16]
USH3	USH3A	*CLRN1*	Clarin-1	Transmembrane protein	[17–19]
n/a	n/a	*PDZD7*	PDZD7	PDZ scaffold protein	[20]

Mutations of *CDH23* and *PCDH15* genes cause nonsyndromic autosomal recessive deafness as well as syndromic deafness (Usher syndrome type I) [4–8]. Usher syndrome (USH) is an autosomal recessive genetic disease, characterized by the association of hearing loss with retinitis pigmentosa and occasional balance problems. USH is classified into three types based on the severity of clinical symptoms. In terms of hearing phenotype, USH type I (USH1) patients suffer from congenital severe-to-profound deafness; USH2 patients suffer from congenital moderate-to-severe hearing loss; USH3 patients, the least severe ones, show progressive hearing loss. Ten USH proteins have been identified so far, including five USH1 proteins, three USH2 proteins, one USH3 protein, and one USH modifier protein [9] (Table 4.1). All known USH proteins could be detected in the hair-cell stereocilia and are able to interact with each other and form protein complexes [21, 22]. USH proteins are required for the development and function of hair bundles, and some of them, especially CDH23 and PCDH15, play an important role in MET.

4.1.1 CDH23 and Upper Tip-Link Complex

CDH23 Gene and Protein

CDH23 is a large single transmembrane protein, which belongs to atypical cadherin. It consists of multiple extracellular cadherin (EC) repeats, a single transmembrane fragment, and a short cytoplasmic domain (Fig. 4.1). *Cdh23* gene contains 69 exons, and transcription starting from different initiation sites gives rise to various transcripts encoding proteins with different numbers of EC repeats [23]. Typical EC repeat is composed of ~110 amino acids that form 7 tightly packed β-strands and adopts a "Greek key" topology that spans approximately 4 nm [24]. The longest CDH23 isoform has 27 EC repeats and hence could span more than 100 nm. In

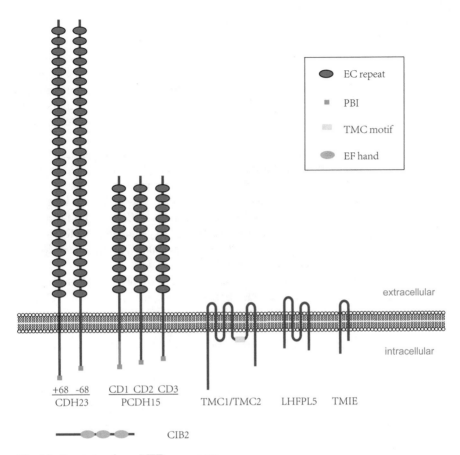

Fig. 4.1 Structures of core MET components

classical cadherins, the linker region between EC repeats binds Ca^{2+} ions, which makes it very rigid [24]. Molecular dynamics simulations suggested that the EC repeats of CDH23 are also quite stiff [25].

The cytoplasmic part of CDH23 bears no homology with that of classical cadherins. As a result, it does not interact with β-catenin as classical cadherins do. CDH23 contains a type-I PDZ domain-binding interface (PBI) at its C-terminal end, which mediates the interaction with PDZ domains [26–29]. Exon 68 of *Cdh23* gene encodes part of the cytoplasmic domain and is subjected to alternative splicing [30]. The resulting *Cdh23* variants have different expression patterns: *Cdh23(−68)* is expressed broadly in various tissues, whereas *Cdh23(+68)* is preferentially expressed in the inner ear [1, 28]. The biological significance of the inclusion or exclusion of exon 68 has not been fully understood yet, although it might affect the conformation of CDH23 and protein-protein interaction [27, 29, 31].

CDH23 Is an Upper Tip-Link Component

The first evidence of CDH23 as a tip-link component came from its localization near the tips of stereocilia [1]. Immunostaining with specific antibodies localized CDH23 on the stereocilia, at a position corresponding to the upper end of tip links [3]. CDH23 was also associated with transient lateral links as well as kinociliary links, both of which are important extracellular links during hair bundle development [32]. Moreover, the recombinant extracellular fragment of CDH23 formed coiled homodimers, in which the membrane-proximal C-terminus separated into two strands [3], just as the upper insertion site of tip links did [33]. CDH23 interacts with the lower tip-link component PCDH15 via their most N-terminal two EC repeats, forming a CDH23-PCDH15 heterotetramer that is about 180 nm long, consistent with the length of tip links [3, 34]. As mentioned in Chap. 2, tip links were disrupted by application of Ca^{2+} chelators such as EGTA, while exogenous application of CDH23 fragments encompassing the extracellular EC1 inhibited tip-link recovery from EGTA-induced disruption [35].

If CDH23 is a component of tip links, inactivation of *Cdh23* gene in mice would cause tip-link disruption as well as MET deficit. Indeed, in *waltzer* mice, a *Cdh23*-null mutant mouse line, tip links were missing, and MET was affected, consistent with the role of CDH23 as a tip-link component [36]. However, in the *Cdh23*-null mice, the development of hair bundle was also severely affected, and the stereocilia were never developed into a mature morphology [37]; hence it's difficult to separate the role of CDH23 in hair bundle development from tip-link formation and MET. The final conclusive proof came from a mutant mouse line named *salsa* that harbours a point mutation in *Cdh23* gene [38]. In *salsa* mice, hair bundles were developed normally, and tip links were formed during early developmental stages and then started to disappear after postnatal day 10 (P10), accompanied by increased hearing threshold [38]. Similar phenotypes were also observed later in another mutant mouse line *jera* that contains different point mutation in *Cdh23* gene [39].

CDH23 and Upper Tip-Link Complex

As mentioned in Chap. 2, the upper tip-link insertion sites on the lateral shaft of taller stereocilia display electron-dense plaques, referred as upper tip-link density (UTLD). Several proteins have been shown to interact with the cytoplasmic domain of CDH23 and participate in the formation of UTLD. USH1C protein harmonin is a scaffold protein containing three PDZ domains, of which PDZ2 binds the C-terminal PBI of CDH23 [26, 27]. Additionally, a highly conserved N-terminal fragment preceding PDZ1 of harmonin binds an internal fragment of CDH23 [40]. Immunostaining showed that harmonin localized at UTLD [41], and this localization was dependent on the presence of CDH23 [42]. Deletion of the coiled-coil and PST domains between harmonin PDZ2 and PDZ3 in deaf circler (*dfcr*) mice prevented the formation of UTLD, suggesting that harmonin plays a central role in UTLD [41]. Harmonin could also bind F-actin; hence it might connect tip links to the cytoskeleton [26].

Two other USH1 proteins, USH1B protein myosin 7a (MYO7A) and USH1G protein SANS, also bind the cytoplasmic domain of CDH23 [42, 43]. Both MYO7A and SANS localized at the UTLD, whereas in *Cdh23*-null mutant mice, their stereociliary localization was altered or completely absent [42, 44]. *Myo7a* mutation in mice resulted in MET deficits and altered adaptation [45], while *sans* inactivation in mice led to loss of tip links and MET deficits [43]. Taken together, CDH23, harmonin, MYO7A, and SANS interact with each other and form the core of upper tip-link complex.

Upper Tip-Link Complex and Slow Adaptation

Slow adaptation occurs on a timescale of several milliseconds. According to the commonly accepted adaptation motor model, slow adaptation is mediated by a motor that sets the MET channel operating point [46, 47]. The motor is associated with tip links as well as the F-actin core near the upper tip-link insertion site and hence regulates the tension of tip links and sets the operating point of the MET channels. In the resting state, the motor climbs up along the stereocilia and hence sets the resting tension of the tip links. When the stereocilia are deflected in the positive direction, tension of tip links increases, and Ca^{2+} enters stereocilia through the MET channels. Ca^{2+} then diffuses to the upper tip-link insertion site and causes the motor to slide down the stereocilia through an unknown mechanism. This movement of motor decreases the tension of tip links and resets the MET channel operating point to the resting level.

Slow adaptation rates vary considerably between vestibular and cochlear hair cells, as well as among different species [46, 48–51]. Hence it is possible that the molecular mechanisms for slow adaptation in vestibular and cochlear hair cells in different species are different. At present, the most prominent candidate of adaptation motor is an unconventional myosin MYO1C. In frogs, MYO1C was concentrated at the UTLD [52, 53]. In mammals, MYO1C was shown to localize along the length of stereocilia [54], and it could interact with the upper tip-link component CDH23 [1]. The most convincing evidence came from a transgenic mouse model that expresses a mutant MYO1C Y61G. The substitution of tyrosine (Y) to glycine (G) in the nucleotide-binding pocket of MYO1C conferred susceptibility to inhibition by N^6-modified ADP analogs such as NMB-ADP [55]. The MYO1C Y61G transgenic mice had normal MET in the absence of NMB-ADP; however, when NMB-ADP was present, slow adaptation in the vestibular hair cells was blocked [56]. Similar results were observed in MYO1C Y61G knockin mice, supporting an important role of MYO1C in slow adaptation [57].

The second candidate for the adaptation motor is MYO7A. As mentioned above, MYO7A is a component of the UTLD and could bind CDH23. Moreover, slow adaptation was affected in *Myo7a* mutant mice [45]. Similarly, another UTLD component harmonin also binds CDH23 and F-actin, and adaptation was significantly slowed in *harmonin* mutant mice *dfcr* [41]. Given their actin-binding ability, MYO1C, MYO7A, or harmonin could act as the adaptation motor. Alternatively, they might affect adaptation through regulating the localization or assembly of the adaptation motor.

4.1.2 PCDH15 and Lower Tip-Link Complex

PCDH15 Is a Lower Tip-Link Component

Similar to CDH23, PCDH15 is also an atypical cadherin. PCDH15 consists of multiple EC repeats, a single transmembrane fragment, and a short cytoplasmic domain (Fig. 4.1). Several lines of evidence supported PCDH15 as a component of tip links. Immunostaining with specific antibodies localized PCDH15 near the tips of shorter stereocilia, a position corresponding to the lower end of tip links [3]. PCDH15 interacts with the upper tip-link component CDH23 via their N-terminal two EC repeats, and the resulting PCDH15-CDH23 heterotetramer is about 180 nm long, consistent with the length of tip links [3, 34]. Exogenous application of the extracellular fragments of PCDH15 inhibited tip-link recovery from EGTA-induced disruption [35].

Further evidences came from *Pcdh15*-null or mutant mice. Tip links were substantially reduced in early postnatal cochlear hair cells of *Pcdh15*-null mutant mice *Ames waltzer 3J (av3J)* [36]. Comparatively, tip links were reduced to a less extent in early postnatal cochlear hair cells of *Pcdh15*-mutant mice *av6J*, in which part of PCDH15 EC9 is deleted [36]. MET currents were reduced in both *av3J* and *av6J* mice [36]. *Pcdh15*-mutant mice *noddy* harbours an isoleucine-to-asparagine (I108N) mutation in EC1 repeat, which impairs the interaction between PCDH15 and CDH23. In *noddy* mice, both bundle morphology and tip-link formation were disrupted, and MET in cochlear hair cells was also affected [58].

PCDH15-CD2 Is Indispensable for Tip-Link Formation

Pcdh15 gene contains 39 exons, and alternative splicing produces various PCDH15 isoforms with different numbers of EC repeats. The longest PCDH15 isoform contains 11 EC repeats. Different from classical cadherins, some linker regions between adjacent PCDH15 EC repeats are calcium-free or partially calcium-free, which might confer some elasticity to the otherwise rigid tip links [59, 60].

Furthermore, according to the alternative inclusion of exon 35, 38, or 39, *Pcdh15* transcript variants are classified into three groups: *Pcdh15-CD1*, *Pcdh15-CD2*, and *Pcdh15-CD3*, respectively [2]. PCDH15-CD1, PCDH15-CD2, and PCDH15-CD3 differ in their C-terminal cytoplasmic domains except for a common region (CR) that is proximal to the transmembrane fragment. Nevertheless, they all contain a type-I PBI at their C-terminal end, which mediates interaction with PDZ domains [21, 61, 62]. The alternative inclusion of exon 26a gives rise to the fourth PCDH15 group without a transmembrane domain and hence might represent a secretory protein and is not likely involved in the formation of tip links.

Expression of different PCDH15 isoforms was examined by performing immunostaining, which showed that they have different spatiotemporal expression patterns in the developing and mature inner ear [2]. In the developing organ of Corti, PCDH15-CD2 immunoreactivity was stronger at the apical turn, whereas PCDH15-CD1 and PCDH15-CD3 immunoreactivity was more intense at the basal turn. PCDH15-CD1 was distributed evenly along the length of the stereocilia except for the tips in mature cochlear hair cells. PCDH15-CD3 immunoreactivity detected with one antibody is restricted to the tips of the shorter stereocilia in immature IHCs and OHCs, but not present in the mature hair cells. Another antibody, however, detected PCDH15-CD3 at the tips of all the stereocilia in adult OHCs, but not in IHCs. In the same study, PCDH15-CD2 immunoreactivity was shown to decrease to an undetectable level in mature cochlear stereocilia [2]. However, another study showed that PCDH15-CD2 immunoreactivity localized at the tips of the three rows of stereocilia in both immature and mature IHCs and OHCs [63]. These discrepancies might result from variations in epitopes recognized by different antibodies or posttranslational modifications of the epitopes.

Knockout mice that lack each of the three PCDH15 isoforms (*Pcdh15-ΔCD1*, *Pcdh15-ΔCD2*, and *Pcdh15-ΔCD3*) helped us to learn more about the role of each isoforms [64]. Hair bundle development and auditory function were normal in *Pcdh15-ΔCD1* and *Pcdh15-ΔCD3* mice. However, in *Pcdh15-ΔCD2* mice, kinociliary links were lost, and hair bundles were misoriented, which contributed to profound hearing loss. Nevertheless, tip links were still present in P1 hair cells in all three knockout mice, suggesting that none of the three PCDH15 isoforms is indispensable for tip-link formation in immature hair bundles. To circumvent the functional redundancy among these PCDH15 isoforms in tip-link formation during early development, conditional *Pcdh15-CD2* knockout mice were developed so that the CD2-encoded exon 38 was deleted only in adult hair cells [63]. In the conditional *Pcdh15-CD2* knockout mice, hair bundles were correctly oriented, and the auditory function was normal by P15. But by P30, the conditional *Pcdh15-CD2* knockout mice were profoundly deaf, and tip links were almost completely lost, suggesting that PCDH15-CD2 is essential for tip-link formation in mature auditory hair cells. In line with this, MET was reduced when a dominant negative PCDH15-CD2 fragment was expressed in wild-type hair cells [65].

PCDH15 and Lower Tip-Link Complex

The lower tip-link insertion site at the tips of shorter stereocilia displays electron-dense plaques, referred as lower tip-link density (LTLD). Four transmembrane proteins have been reported to localize at the lower tip-link insertion site, which are TMC1, TMC2, LHFPL5, and TMIE. These transmembrane proteins interact with PCDH15 and are considered as integral components of MET machinery. These proteins will be discussed in more details in the following sessions.

LTLD might also involve cytosolic proteins that interact with PCDH15 cytoplasmic domain. SANS interacts with PCDH15-CD2 and PCDH15-CD3, but not PCDH15-CD1, and has been located at the tips of short and middle row stereocilia, close to LTLD [43]. Harmonin and MYO7A bind the cytoplasmic domain of PCDH15-CD1 [21, 61, 66]; however their localization at LTLD has not been reported.

4.2   Mechanotransduction Machinery

Identification of the MET machinery components, especially the pore-forming channels, has been a great challenge because of the scarcity of hair cells and the low expression levels of the channels and associated proteins. For example, in the mouse cochlea, there are about 15,000 hair cells with around 100 functional tip links within each hair cell, and only one or two functional channels are estimated to be associated with each tip link [67, 68]. Genetic studies have suggested several candidates for MET channels, most of which were disapproved later on. Recently, many lines of evidence suggested that four integral transmembrane proteins, TMC1, TMC2, LHFPL5, and TMIE, as well as a soluble protein CIB2, might form the core component of MET machinery. Especially, TMC1 and TMC2 were proposed as the pore-forming subunits of the MET channel. However, further investigation is needed before a final conclusion could be drawn.

4.2.1   TMC1/TMC2

TMC1/TMC2 Gene and Protein

Transmembrane channel-like 1 and 2 (TMC1 and TMC2) are homologous proteins with six predicted transmembrane domains and two additional hydrophobic domains that are not predicted to span the membrane [69]. TMC1 and TMC2 are members of TMC protein family that includes eight proteins in humans or mice [70, 71]. Mouse *Tmc1* gene contains 24 exons, and exon 2 is subjected to alternative splicing [72]. The two *Tmc1* variants use different translation initiation codons in exons 1 and 2, respectively, giving rise to TMC1 protein with different N-termini. Both splicing variants were detected in the mouse inner ear with similar expression levels [72].

In mammals, eight TMC proteins (TMC1 through 8) form a distinct protein family, sharing six-transmembrane domains as well as a highly conserved TMC motif with unknown function [70] (Fig. 4.1). The topology of TMC1 has been investigated by inserting HA epitope tag at different positions of heterogeneously expressed TMC1. The result showed that TMC1 is an integral membrane protein localized at ER, containing six-transmembrane domains and cytosolic N- and C-termini [73].

Heterogeneously expressed TMC1 or TMC2 is unsuccessful to be targeted to the plasma membrane at present, which hinders further functional examination of these proteins.

In situ hybridization revealed that *Tmc1* and *Tmc2* were expressed specifically in auditory and vestibular hair cells in the mouse inner ear [72]. This hair cell-specific expression of *Tmc1/Tmc2* was further supported by RNA-seq results (SHIELD; https://shield.hms.harvard.edu) [74, 75]. The temporal expression profile of *Tmc1/ Tmc2* in the mouse inner ear was examined by quantitative RT-PCR (qPCR) analysis [72]. In the cochlea, *Tmc2* expression started at P0–P2, around the same time that hair cells became mechanosensitive (P1–P3) [76]. Then *Tmc2* expression declined at P3–P5, while *Tmc1* expression started. In the adult mouse cochlea, *Tmc2* expression dropped to an undetectable level, whereas *Tmc1* expression persisted [72]. Both *Tmc1* and *Tmc2* were expressed in the vestibular system through development to adulthood [72].

TMC1/TMC2 Localizes Near the Lower End of Tip Links and Binds PCDH15

The subcellular localization of TMC1 and TMC2 in hair cells was examined using BAC-transgenic mice that express fluorescence protein-tagged TMC1 and TMC2. The result showed that both TMC1 and TMC2 were concentrated near the lower end of tip links [77], where the transduction channels were suggested to localize [68]. Similar results were obtained with antibodies against the endogenous TMC1 and TMC2 [77]. This localization puts TMC1/TMC2 at the centre of the MET machinery.

Moreover, immunoprecipitation experiments showed that heterogeneously expressed TMC1 and TMC2 interact with the lower tip-link component PCDH15, which further supported the role of TMC1/TMC2 as an integral component of MET machinery [78, 79]. However, caution is needed to be exercised in interpreting these results, since when expressed heterogeneously, TMC1/TMC2 and PCDH15 were distributed in different intracellular compartments.

TMC1 Disruption Causes Hearing Loss

TMC1 gene mutations are responsible for recessive and dominant nonsyndromic autosomal deafness, DFNB7/B11 and DFNA36, respectively [69]. Meanwhile, mutations of *Tmc1* gene cause recessive hearing loss in *deafness* (*dn*) mice as well as semidominant hearing loss in *Beethoven* (*Bth*) mice [69, 80]. *Dn* mice were originally identified as spontaneous mutant mice that are congenitally deaf [81]. *Dn* mice contains an in-frame deletion of exon 14 (171 bp) of *Tmc1* gene [69] and showed significant hair-cell loss by P30 [82]. *Bth* mice harbour a methionine-to-lysine (M412 K) point mutation in *Tmc1* gene. In *Bth/+* mice, progressive hair-cell degeneration started from P20, followed by progressive hearing loss from around P30. In

Bth/Bth mice, hair-cell degeneration was much more severe, and *Bth/Bth* mice were congenitally deaf [80]. Mutations in human or mouse *TMC2/Tmc2* gene have not been reported so far. Consistently, hearing threshold was normal in *Tmc2* knockout mice [72].

TMC1/TMC2 Disruption Affects MET

Consistent with the fact that *Tmc2* knockout mice had normal hearing threshold, the MET currents were normal in early postnatal auditory hair cells of *Tmc2* knockout mice [72]. Surprisingly, despite the fact that homozygous *dn* and *Bth* mice were congenitally deaf, the MET currents were normal in early postnatal auditory hair cells of these *Tmc1* mutant mice [82]. Similar result was obtained in *Tmc1* knockout mice [72]. However, the MET currents were completely absent in *Tmc1/Tmc2* double knockout mice; meanwhile, hair bundle morphology and tip links were unaffected [72, 83, 84]. The specific deficit of MET in *Tmc1/Tmc2* double knockout mice suggested that TMC1 and TMC2 are important components of the MET machinery.

Transportation of TMC1/TMC2 to the Stereocilia Requires TOMT/ LRTOMT

Transportation of TMC1/TMC2 to the tips of stereocilia is tightly regulated by other proteins, one of which has recently been identified as transmembrane O-methyltransferase (TOMT) [85, 86]. The human ortholog of *Tomt* gene is called *leucine rich transmembrane and O-methyltransferase domain containing* (*LRTOMT*), which has evolved from the fusion of *TOMT* gene with the neighbouring *LRRC51* gene, and mutations in *LRTMOT* gene cause nonsyndromic recessive deafness DFNB63 [87, 88]. Similarly, mutations in mouse or zebrafish *TOMT* gene cause progressive degeneration of hair cells and profound hearing loss [85, 86, 88]. Further investigation revealed that MET currents were completely absent in *TOMT*-deficient mice and zebrafish [85, 86].

The subcellular localization of TOMT in mouse hair cells was examined by immunostaining, which showed that endogenous TOMT was distributed throughout the cytoplasm of hair cells, but not in the stereocilia [87]. Similar results were obtained when epitope-tagged TOMT/LRTOMT was expressed in transgenic zebrafish or cultured mouse cochlear sensory epithelium [85, 86]. Epitope-tagged TOMT/LRTOMT partially co-localized with the Golgi marker in hair cells of transgenic zebrafish, or ER marker in heterologous cells, suggesting a possible role in the regulation of membrane protein transportation [85, 86]. In line with this, epitope-tagged TOMT was co-immunoprecipitated with TMC1/2, LHFPL5, TMIE, and PCDH15-CD2 in vitro, and TMC1/2 was absent in the stereocilia in *TOMT*-deficient mice or zebrafish [85, 86]. Taken together, the present data suggested that TOMT plays an essential regulatory role in the transportation of TMC1/2 to the stereocilia and is indispensable for hair-cell MET. Meanwhile, despite of the important role of TOMT in TMC1/2 transportation, co-expression of TOMT did not alter the cytoplasmic localization of TMC1/2 in heterologous cells, suggesting that protein(s) other than TOMT is also necessary to target TMC1/2 to the plasma membrane [85].

Are TMC1/TMC2 the MET Channels?

As mentioned above, many lines of evidence are consistent with the hypothesis that TMC1/TMC2 are MET channels of mammalian auditory hair cells. First, TMC1 and TMC2 are expressed during the right time and at the right place. In the mouse cochlea, hair cells become mechanosensitive at P1–P3, whereas *Tmc2* expression starts at P0–P2, followed by expression of *Tmc1*. In the mouse cochlea, *Tmc1* and *Tmc2* are exclusively expressed in hair cells. Within hair cells, TMC1 and TMC2 were shown to localize near the tips of shorter stereocilia, coincident with the localization of MET channels. Second, heterologously expressed TMC1/TMC2 could be co-immunoprecipitated together with the lower tip-link component PCDH15, which also localizes at the tips of shorter stereocilia. Third, *Tmc1/Tmc2* double knockout completely eliminates the MET currents in mouse hair cells while leaving hair bundle and tip links unaffected.

At present the role of TMC1/TMC2 as MET channels is under hot debate [89–91]. Other proteins such as LHFPL5 and TMIE behave similarly to TMC1/TMC2: they localize at the tips of shorter stereocilia and interact with PCDH15, and their mutations lead to deafness as well as loss of MET current (discussed below). Probably the biggest challenge comes from the fact that TMC1 or TMC2 so far has not been successfully shown to serve as a channel in a heterologous expression system. Heterogeneously expressed TMC1 localizes at the ER, not on the plasma membrane [73], which hinders further examination of its potential channel activity. Heterologously expressed TMC-1, one of the two TMCs that exist in *C. elegans*, was reported to have sodium channel activity [92], whereas similar observation has not been reported on mammalian TMCs yet.

4.2.2 LHFPL5

LHFPL5 Gene and Protein

LHFPL5 (lipoma HMGIC fusion partner-like 5), also known as TMHS (tetraspan membrane protein of hair-cell stereocilia), is a predicted tetraspan transmembrane stereociliary protein, as its name implicates. Human *LHFPL5* gene contains 4 exons, encoding 219 amino acids. *LacZ* reporter assay and in situ hybridization showed that *Lhfpl5* mRNA was expressed in mouse cochlear and vestibular hair cells [93, 94]. The hair cell-specific expression of *Lhfpl5* was further supported by RNA-seq results (SHIELD; https://shield.hms.harvard.edu) [74, 75].

LHFPL5 is predicted to contain four transmembrane helices as well as two extra-cellular loops (Fig. 4.1). LHFPL5 belongs to a small tetraspan transmembrane protein family that includes LHFP as well as LHFP-like 1–5 (LHFPL1–5) [95, 96], whose biological functions largely remain unknown. In general, tetraspan proteins constitute a large protein superfamily, which includes claudin tight junction proteins, connexin gap junction proteins, clarins, etc. Some tetraspan proteins are involved in channel activity. For example, some claudins act as channels that specifically allow certain ions to cross tight junctions [97]. Connexins are subunits of gap junctions that form transmembrane channels connecting the cytoplasm of adjacent cells [98]. LRRC8 family tetraspan proteins form the pore of volume-regulated anion channels (VRAC) that sense ionic strength [99]. Additionally, tetraspan proteins TARPs regulate AMPA channel activity as auxiliary subunits [100]. It awaits further investigation whether LHFPL5 could act as pore-forming subunit or auxiliary subunit of channels.

LHFPL5 Localizes Near the Lower End of Tip Links and Binds PCDH15

Immunostaining showed that LHFPL5 localized near the lower end of tip links in hair cells [94], putting LHFPL5 at the position where MET machinery localizes [68]. Moreover, LHFPL5 directly binds to the transmembrane and membrane-proximal domains of PCDH15 [94]. Further investigation demonstrated that LHFPL5 was necessary for the localization of PCDH15 and TMC1 on the stereocilia, thereby controlling the formation of tip link and MET machinery [79, 94].

LHFPL5 Disruption Causes Hearing Loss

Mutations in *LHFPL5* gene cause autosomal recessive nonsyndromic deafness DFNB67 [101, 102]. *Lhfpl5* mutation is also responsible for profound hearing loss and balance deficits in *hurry-scurry* (*hscy*) mice [96]. *Hscy* mice harbour a cysteine-to-phenylalanine (C161F) missense mutation in LHFPL5, which occurs in the second extracellular loop. The mutant LHFPL5 is unstable and is undetected in the inner ear of *hscy* mice. In *hscy* mice, stereocilia were disorganized when examined at P8, followed by hair-cell degeneration [96]. *Lhfpl5* knockout mice were also developed and showed phenotype identical to that of the *hscy* mice [93].

LHFPL5 Disruption Affects MET

Recently, *Lhfpl5* mutation was shown to lead to a nearly 90% reduction in MET in cochlear hair cells of *hscy* mice and *Lhfpl5* knockout mice [94]. Single-channel recordings revealed that the conductance of MET was reduced and adaptation was severely impaired in the absence of LHFPL5, suggesting that LHFPL5 is an integral component of the MET machinery [94]. Moreover, the tonotopic gradient in the conductance of the transducer channels was also blunted in *Lhfpl5* knockout mice [79].

4.2.3   TMIE

TMIE Gene and Protein

Human *TMIE* gene consists of 6 exons that encode a small protein of 156 amino acids. Northern blot showed that three *Tmie* transcripts were expressed in various mouse tissues with a molecular weight of 2.2 kb, 2.8 kb, and 3.2 kb, respectively [103]. RNA sequencing revealed that *Tmie* was enriched in hair cells in mouse organ of Corti (SHIELD; https://shield.hms.harvard.edu) [74, 75]. *LacZ* reporter assay also showed that *Tmie* mRNA was specifically expressed in cochlear and vestibular hair cells [65].

TMIE contains two predicted transmembrane domains (Fig. 4.1). Consistently, heterologously expressed TMIE localized on the plasma membrane of HEK293 cells [65, 104]. Western blot with a polyclonal antibody detected a single band of approximately 17 kD in rat tissues [105]. TMIE shows no homology with any known proteins, and at present its molecular function remains elusive. It is worth noticing that proteins with two transmembrane domains such as ENaC/DEG have been shown to form the pore of MET channels in *C. elegans* [106].

TMIE Localizes Near the Lower End of Tip Links and Binds PCDH15

Immunostaining revealed that TMIE localized at the tips of shorter stereocilia, where the lower end of tip links inserts [65]. In cultured cochlear explants that injectoporated with TMIE-HA expression plasmid, HA immunoreactivity accumulated at the tips of the shorter rows of stereocilia, confirming the localization of TMIE near the lower end of tip links [65]. Furthermore, yeast two-hybrid screen and co-immunoprecipitation experiments showed that TMIE directly binds LHFPL5 as well as PCDH15-CD2 [65].

The localization of TMIE and its interaction with known MET components suggested that it is an integral component of hair-cell MET machinery. TMIE, LHFPL5, and PCDH15 could form a ternary complex, but the geometry of this protein complex differs when different PCDH15 isoforms are present. When PCDH15-CD2 is present, TMIE, LHFPL5, and PCDH15-CD2 bind to one another directly. On the other hand, when PCDH15-CD1 or PCDH15-CD3 is present, TMIE and PCDH15-CD1/CD3 bind to each other indirectly via LHFPL5. The TMIE/LHFPL5/PCDH15 complex might further connect to TMC1/TMC2 via PCDH15. Hence LHFPL5 seems to play a central role in the organization of this protein complex. Consistently, disruption of LHFPL5, but not TMIE, affected tip-link assembly as well as the expression of TMC1/TMC2 on stereocilia [65, 79, 94].

TMIE Disruption Causes Hearing Loss

Mutations of *TMIE* cause autosomal recessive nonsyndromic deafness DFNB6 [107]. *Tmie* mutations are also responsible for hearing loss and vestibular dysfunction in *spinner (sr)* as well as *circling (cir)* mutant mice [103, 108]. *Sr* and *cir* mice have deletions in the genome that include the entire *Tmie* gene, while *sr^J* mice contain a single-base-pair substitution that truncates TMIE protein. Both *sr* and *cir* mice showed disorganized stereocilia of auditory hair cells after P10 [103, 109]. Similarly, when selective exons of *Tmie* gene were deleted through homologous recombination, the mice showed profound hearing deficits [65].

TMIE Disruption Affects MET

MET was completely abolished in hair cells of *Tmie* knockout mice [65]. Moreover, MET was reduced when a dominant negative TMIE fragment was expressed in wild-type hair cells [65]. These results suggested that TIME is indispensable for MET. Surprisingly, although TMIE disruption completely abolished MET in hair cells, it didn't affect the localization of other known components of the MET machinery in the stereocilia [65]. The mechanism by which TMIE affects MET awaits further investigation.

4.2.4 CIB2

CIB2 Gene and Protein

Human *CIB2* (*calcium- and integrin-binding protein 2*) gene contains 6 exons encoding 3 isoforms, the longest of which consists of 210 amino acids [110, 111]. *Cib2* was expressed ubiquitously in multiple tissues such as skeletal muscle, the brain, the eye, and the inner ear [110–112]. RNA sequencing revealed that *Cib2* was enriched in mouse auditory and vestibular hair cells (SHIELD; https://shield.hms.harvard.edu) [74, 75]. Consistently, *LacZ* reporter assay showed that *Cib2* mRNA was specifically expressed in hair cells in mouse inner ear [113].

Different from the MET components discussed above, CIB2 is a soluble protein. CIB2 belongs to a protein family that includes CIB1 through CIB4, which is characterized by multiple calcium-binding EF-hand domains [114]. CIB2 contains three EF-hand domains and binds calcium through the last two EF-hand domains [115] (Fig. 4.1). Moreover, CIB2 underwent N-myristoylation and was associated with intracellular membranes in neurons, co-localizing with Golgi apparatus and dendrite markers [115]. Known CIB2-binding partners include integrins αIIb and α7b, myosin VIIA, whirlin, and sphingosine kinase 1 (SK1) [111, 112, 116, 117]. CIB2 has been suggested to regulate HIV-1 entry through affecting the surface receptor such as CXCR4, CCR5, and integrin α4β7, but the detailed mechanism remains elusive [118, 119].

CIB2 Is Concentrated Near the Tips of Stereocilia and Interacts with TMC1/TMC2

In the inner ear, CIB2 immunoreactivity localized along the length of hair-cell stereocilia and was concentrated at the tips of stereocilia [111, 113]. Concentrated localization at the stereocilia tips was also observed when CIB2-GFP was transfected into auditory or vestibular hair cells using gene gun [111]. Notably, both endogenous and exogenous CIB2 were more concentrated at the tips of shorter row stereocilia [111, 113]. This localization suggested that CIB2 might associate with the MET machinery. Consistent with this hypothesis, fluorescence resonance energy transfer (FRET) and co-immunoprecipitation experiments showed that CIB2 interacts with TMC1/TMC2 in vitro, and the interaction was affected by deafness-associated CIB2 mutations [113]. However, the localization of MET components TMC1/2 and PCDH15 on stereocilia was not affected by CIB2 deficiency, suggesting that CIB2 is not essential for transporting or stabilizing these proteins to stereocilia [113].

CIB2 Disruption Causes Hearing Loss

Mutations in the *CIB2* gene are associated with nonsyndromic hearing loss DFNB48 [111, 120, 121]. *CIB2* mutation has also been reported to lead to syndromic hearing loss USH1J [111], but recently it was suggested that CIB2 mutation might not cause Usher syndrome [122, 123]. Morpholino knockdown of *Cib2* expression in the zebrafish embryo resulted in reduced or even absent response to acoustic stimuli as well as balancing problems [111]. To further explore the role of CIB2 in hearing, several lines of *Cib2*-deficient mice have been established, including $Cib2^{tm1a}/Cib2^{tm1b}$, $Cib2^{F91S}$, $Cib2^{\Delta ex4}$, and $Cib2^{\Delta 17bp}$. $Cib2^{tm1a}$ mice were generated by inserting a gene trap cassette containing *lacZ* and neomycin resistance genes between *Cib2* exons 3 and 4 [113, 124]. $Cib2^{tm1b}$ mice were obtained by crossing $Cib2^{tm1a}$ mice with Cre-expressing mice to delete the neomycin cassette and exon 4 of *Cib2* [113]. $Cib2^{F91S}$ knockin mice carry a p.F91S missense mutation, which is the most prevalent *CIB2* mutation that causes nonsyndromic deafness [113]. $Cib2^{\Delta ex4}$ mice were generated by putting loxP sites flanking exon 4 of *Cib2* gene and crossing with Cre-expressing mice to delete exon 4 [122]. Lastly, $Cib2^{\Delta 17bp}$ mice were generated using CRISPR/Cas9 technique that frameshift deletions (8 and 9 bp, respectively) were introduced into exon 4 of *Cib2* gene [125].

All the *Cib2*-deficient mice showed profound hearing loss, confirming that CIB2 plays an important role in hearing transduction [113, 122, 124, 125]. Interestingly, the shorter row stereocilia in *Cib2*-deficient OHCs and IHCs were over-elongated, whereas the tallest stereocilia remained unaffected [113, 125]. Furthermore, the kinocilia in *Cib2*-deficient IHCs do not regress properly during development [113, 122, 125].

CIB2 Disruption Affects MET

FM1-43 dye uptake and microphonic potential were reduced in the lateral-line hair cells of *Cib2* zebrafish morphants, suggesting that MET is affected by *Cib2* knock-down [111]. In *Cib2^tm1a^* hair cells, FM1-43 dye uptake was even completely abolished [113]. Consistently, whole-cell patch-clamp recordings showed that the conventional MET currents in *Cib2*-deficient IHCs and OHCs were completely absent, whereas the reverse-polarity MET was unaffected [113, 122, 125]. Taken together, these results suggest that CIB2 plays an indispensable role in hair-cell MET.

4.2.5 PIEZO2

In mature hair cells, deflection of the hair bundles towards the tallest stereocilia increases the open probability of the sensory MET channels, while deflection in the opposite direction decreases the open probability [126]. However, in the developing immature hair cells, the hair bundles are less directionally sensitive, and transducer currents can be evoked by deflection of the hair bundles in both directions [127, 128]. Additionally, reverse-polarity currents can also be evoked in hair cells lacking tip links, as well as in hair cells deficient for MYO7A, MYO15A, ADGRV1, TMC1/TMC2, LHFPL5, or TMIE [36, 65, 79, 83, 128–130]. The ion selectivity and responsiveness to pharmacological blockers of the reverse-polarity current are similar but not identical to that of the regular MET current [83, 131]. High-speed Ca^{2+} imaging suggested that the reverse-polarity channels are not localized to the hair bundle but distributed at the apical surface of hair cells [132]. These data suggest that the reverse-polarity MET currents are mediated by channels different from the regular MET channels associated with the lower end of tip links.

Recently, PIEZO2 was suggested to constitute the MET channel responsible for the reverse-polarity currents in mouse hair cells [133]. PIEZO2 and its close homolog PIEZO1 are the first mammalian mechanosensitive ion channels identified so far [134]. Both proteins consist of more than 2500 amino acids that are predicted to encompass 26–40 transmembrane domains. PIEZO2 mediated MET in mouse Merkel cells, dorsal root ganglion cells, proprioceptors, and airway-innervating sensory neurons, and deletion of *Piezo2* gene caused loss of touch sensation and proprioception as well as lung inflation-induced apnoea [135–138]. Mutations of human *PIEZO2* gene are associated with distal arthrogryposis, muscular atrophy, or loss of discriminative touch perception [139–142].

In the mouse inner ear, *Piezo2* expression was detected in OHCs and vestibular hair cells, but not in IHCs. Immunostaining revealed that PIEZO2 localized at the apical surface of OHCs near the tallest stereocilia [133]. In conditional *Piezo2* knockout mice that *Piezo2* was inactivated in the inner ear, the patterning of organ of Corti as well as hair bundle morphology was not affected, whereas OHC function was mildly affected, but no obvious vestibular defect was observed [133]. Interestingly, *Piezo2* disruption did not affect regular hair-cell MET but abolished the reverse-polarity currents in hair cells lacking tip links or in immature hair cells [133]. These data suggest that PIEZO2 is responsible for the reverse-polarity currents in mouse hair cells.

4.3 Discussion

The mechanism of mammalian hair-cell MET has been under intense investigation in the recent several decades, and significant advances have been achieved. At present, researchers have identified the proteins that constitute the tip links as well as other MET components and are on the way to discover the MET channel itself. However, as we pointed out in the above sections, there is still a long way to go before we can fully understand how MET happens in hair cells.

Tip links play a pivotal role in hair-cell MET. Convincing evidences have been provided suggesting that PCDH15 and CDH23 bind each other via their N-termini and constitute the lower and upper component of tip links, respectively. It's worth noting that tip links might not always be composed of PCDH15 and CDH23. For example, during regeneration of disrupted tip links, shorter tip links containing only PCDH15 appeared first and then were replaced by mature PCDH15/CDH23 tip links [143]. It also remains elusive whether tip links could act as the gating spring during MET. High-resolution electron microscopy suggested that tip links are relatively stiff and inextensible [33]. Molecular dynamics simulations based on the structure of the CDH23 EC1–EC2 fragment suggested that the EC repeats of CDH23 are quite stiff [25]. Recently, however, some of the PCDH15 EC repeat linker regions were shown to be calcium-free or partially calcium-free, which might confer some elasticity to the otherwise rigid tip links [59, 60].

PCDH15 and CDH23 interact with other proteins via their transmembrane domains or cytoplasmic domains and form the so-called lower tip-link complex and upper tip-link complex, respectively. The molecular composition of these complexes is just emerging. MYO7A, harmonin, SANS, and possibly MYO1C are considered as components of upper tip-link complex. The adaption motor is positioned near the upper tip-link complex. Through moving along the stereocilia, the motor regulates the tension of tip links and adjusts the operating point of the MET channels. MYO1C is the most prominent candidate for the adaptation motor, whereas MYO7A and harmonin are also suggested to regulate slow adaptation directly or indirectly.

So far four transmembrane proteins have been localized near the lower tip-link insertion site, namely, TMC1, TMC2, LHFPL5, and TMIE. They all interact with PCDH15 in vitro and are likely part of the lower tip-link complex. TMC1 and TMC2 are potential six-transmembrane proteins and have been proposed as pore-forming subunits of hair-cell MET channel, although compelling evidence is needed to fully support this conclusion. LHFPL5 and TMIE contain four and two predicted transmembrane domains, respectively, and are suggested to be integral components of hair-cell MET machinery. Another candidate MET component is CIB2, a soluble protein that is concentrated at the tips of shorter row stereocilia and binds TMC1/TMC2. Characterization of the MET channel surely is the most exiting task in this field in the coming several years.

References

1. Siemens, J., et al. 2004. Cadherin 23 is a component of the tip link in hair-cell stereocilia. *Nature* 428 (6986): 950–955.
2. Ahmed, Z.M., et al. 2006. The tip-link antigen, a protein associated with the transduction complex of sensory hair cells, is protocadherin-15. *The Journal of Neuroscience* 26 (26): 7022–7034.
3. Kazmierczak, P., et al. 2007. Cadherin 23 and protocadherin 15 interact to form tip-link filaments in sensory hair cells. *Nature* 449 (7158): 87–91.
4. Bolz, H., et al. 2001. Mutation of CDH23, encoding a new member of the cadherin gene family, causes Usher syndrome type 1D. *Nature Genetics* 27 (1): 108–112.
5. Bork, J.M., et al. 2001. Usher syndrome 1D and nonsyndromic autosomal recessive deafness DFNB12 are caused by allelic mutations of the novel cadherin-like gene CDH23. *American Journal of Human Genetics* 68 (1): 26–37.
6. Ahmed, Z.M., et al. 2001. Mutations of the protocadherin gene PCDH15 cause Usher syndrome type 1F. *American Journal of Human Genetics* 69 (1): 25–34.
7. Alagramam, K.N., et al. 2001. Mutations in the novel protocadherin PCDH15 cause Usher syndrome type 1F. *Human Molecular Genetics* 10 (16): 1709–1718.
8. Ahmed, Z.M., et al. 2003. PCDH15 is expressed in the neurosensory epithelium of the eye and ear and mutant alleles are responsible for both USH1F and DFNB23. *Human Molecular Genetics* 12 (24): 3215–3223.
9. Mathur, P., and J. Yang. 2015. Usher syndrome: hearing loss, retinal degeneration and associated abnormalities. *Biochimica et Biophysica Acta* 1852 (3): 406–420.
10. Weil, D., et al. 1995. Defective myosin VIIA gene responsible for Usher syndrome type 1B. *Nature* 374 (6517): 60–61.
11. Verpy, E., et al. 2000. A defect in harmonin, a PDZ domain-containing protein expressed in the inner ear sensory hair cells, underlies Usher syndrome type 1C. *Nature Genetics* 26 (1): 51–55.
12. Bitner-Glindzicz, M., et al. 2000. A recessive contiguous gene deletion causing infantile hyperinsulinism, enteropathy and deafness identifies the Usher type 1C gene. *Nature Genetics* 26 (1): 56–60.
13. Weil, D., et al. 2003. Usher syndrome type I G (USH1G) is caused by mutations in the gene encoding SANS, a protein that associates with the USH1C protein, harmonin. *Human Molecular Genetics* 12 (5): 463–471.
14. Eudy, J.D., et al. 1998. Mutation of a gene encoding a protein with extracellular matrix motifs in Usher syndrome type IIa. *Science* 280 (5370): 1753–1757.

15. Weston, M.D., et al. 2004. Mutations in the VLGR1 gene implicate G-protein signaling in the pathogenesis of Usher syndrome type II. *American Journal of Human Genetics* 74 (2): 357–366.
16. Ebermann, I., et al. 2007. A novel gene for Usher syndrome type 2: mutations in the long isoform of whirlin are associated with retinitis pigmentosa and sensorineural hearing loss. *Human Genetics* 121 (2): 203–211.
17. Joensuu, T., et al. 2001. Mutations in a novel gene with transmembrane domains underlie Usher syndrome type 3. *American Journal of Human Genetics* 69 (4): 673–684.
18. Fields, R.R., et al. 2002. Usher syndrome type III: revised genomic structure of the USH3 gene and identification of novel mutations. *American Journal of Human Genetics* 71 (3): 607–617.
19. Adato, A., et al. 2002. USH3A transcripts encode clarin-1, a four-transmembrane-domain protein with a possible role in sensory synapses. *European Journal of Human Genetics* 10 (6): 339–350.
20. Ebermann, I., et al. 2010. PDZD7 is a modifier of retinal disease and a contributor to digenic Usher syndrome. *The Journal of Clinical Investigation* 120 (6): 1812–1823.
21. Adato, A., et al. 2005. Interactions in the network of Usher syndrome type 1 proteins. *Human Molecular Genetics* 14 (3): 347–356.
22. Chen, Q., et al. 2014. Whirlin and PDZ domain-containing 7 (PDZD7) proteins are both required to form the quaternary protein complex associated with Usher syndrome type 2. *The Journal of Biological Chemistry* 289 (52): 36070–36088.
23. Lagziel, A., et al. 2005. Spatiotemporal pattern and isoforms of cadherin 23 in wild type and waltzer mice during inner ear hair cell development. *Developmental Biology* 280 (2): 295–306.
24. Boggon, T.J., et al. 2002. C-cadherin ectodomain structure and implications for cell adhesion mechanisms. *Science* 296 (5571): 1308–1313.
25. Sotomayor, M., et al. 2010. Structural determinants of cadherin-23 function in hearing and deafness. *Neuron* 66 (1): 85–100.
26. Boeda, B., et al. 2002. Myosin VIIa, harmonin and cadherin 23, three Usher I gene products that cooperate to shape the sensory hair cell bundle. *The EMBO Journal* 21 (24): 6689–6699.
27. Siemens, J., et al. 2002. The Usher syndrome proteins cadherin 23 and harmonin form a complex by means of PDZ-domain interactions. *Proceedings of the National Academy of Sciences of the United States of America* 99 (23): 14946–14951.
28. Xu, Z., et al. 2008. MAGI-1, a candidate stereociliary scaffolding protein, associates with the tip-link component cadherin 23. *The Journal of Neuroscience* 28 (44): 11269–11276.
29. Xu, Z., K. Oshima, and S. Heller. 2010. PIST regulates the intracellular trafficking and plasma membrane expression of cadherin 23. *BMC Cell Biology* 11: 80.
30. Di Palma, F., R. Pellegrino, and K. Noben-Trauth. 2001. Genomic structure, alternative splice forms and normal and mutant alleles of cadherin 23 (Cdh23). *Gene* 281 (1-2): 31–41.
31. Yonezawa, S., et al. 2008. Redox-dependent structural ambivalence of the cytoplasmic domain in the inner ear-specific cadherin 23 isoform. *Biochemical and Biophysical Research Communications* 366 (1): 92–97.
32. Michel, V., et al. 2005. Cadherin 23 is a component of the transient lateral links in the developing hair bundles of cochlear sensory cells. *Developmental Biology* 280 (2): 281–294.
33. Kachar, B., et al. 2000. High-resolution structure of hair-cell tip links. *Proceedings of the National Academy of Sciences of the United States of America* 97 (24): 13336–13341.
34. Sotomayor, M., et al. 2012. Structure of a force-conveying cadherin bond essential for inner-ear mechanotransduction. *Nature* 492 (7427): 128–132.
35. Lelli, A., et al. 2010. Development and regeneration of sensory transduction in auditory hair cells requires functional interaction between cadherin-23 and protocadherin-15. *The Journal of Neuroscience* 30 (34): 11259–11269.
36. Alagramam, K.N., et al. 2011. Mutations in protocadherin 15 and cadherin 23 affect tip links and mechanotransduction in mammalian sensory hair cells. *PLoS One* 6 (4): e19183.

37. Di Palma, F., et al. 2001. Mutations in Cdh23, encoding a new type of cadherin, cause ste-reocilia disorganization in waltzer, the mouse model for Usher syndrome type 1D. *Nature Genetics* 27 (1): 103–107.

38. Schwander, M., et al. 2009. A mouse model for nonsyndromic deafness (DFNB12) links hearing loss to defects in tip links of mechanosensory hair cells. *Proceedings of the National Academy of Sciences of the United States of America* 106 (13): 5252–5257.

39. Manji, S.S., et al. 2011. An ENU-induced mutation of Cdh23 causes congenital hearing loss, but no vestibular dysfunction, in mice. *The American Journal of Pathology* 179 (2): 903–914.

40. Pan, L., et al. 2009. Assembling stable hair cell tip link complex via multidentate interactions between harmonin and cadherin 23. *Proceedings of the National Academy of Sciences of the United States of America* 106 (14): 5575–5580.

41. Grillet, N., et al. 2009. Harmonin mutations cause mechanotransduction defects in cochlear hair cells. *Neuron* 62 (3): 375–387.

42. Bahloul, A., et al. 2010. Cadherin-23, myosin VIIa and harmonin, encoded by Usher syn-drome type I genes, form a ternary complex and interact with membrane phospholipids. *Human Molecular Genetics* 19 (18): 3557–3565.

43. Caberlotto, E., et al. 2011. Usher type 1G protein sans is a critical component of the tip-link complex, a structure controlling actin polymerization in stereocilia. *Proceedings of the National Academy of Sciences of the United States of America* 108 (14): 5825–5830.

44. Grati, M., and B. Kachar. 2011. Myosin VIIa and sans localization at stereocilia upper tip-link density implicates these Usher syndrome proteins in mechanotransduction. *Proceedings of the National Academy of Sciences of the United States of America* 108 (28): 11476–11481.

45. Kros, C.J., et al. 2002. Reduced climbing and increased slipping adaptation in cochlear hair cells of mice with Myo7a mutations. *Nature Neuroscience* 5 (1): 41–47.

46. Assad, J.A., and D.P. Corey. 1992. An active motor model for adaptation by vertebrate hair cells. *The Journal of Neuroscience* 12 (9): 3291–3309.

47. Yamoah, E.N., and P.G. Gillespie. 1996. Phosphate analogs block adaptation in hair cells by inhibiting adaptation-motor force production. *Neuron* 17 (3): 523–533.

48. Eatock, R.A., D.P. Corey, and A.J. Hudspeth. 1987. Adaptation of mechanoelectrical trans-duction in hair cells of the bullfrog's sacculus. *The Journal of Neuroscience* 7 (9): 2821–2836.

49. Holt, J.R., D.P. Corey, and R.A. Eatock. 1997. Mechanoelectrical transduction and adap-tation in hair cells of the mouse utricle, a low-frequency vestibular organ. *The Journal of Neuroscience* 17 (22): 8739–8748.

50. Wu, Y.C., A.J. Ricci, and R. Fettiplace. 1999. Two components of transducer adaptation in auditory hair cells. *Journal of Neurophysiology* 82 (5): 2171–2181.

51. Stauffer, E.A., and J.R. Holt. 2007. Sensory transduction and adaptation in inner and outer hair cells of the mouse auditory system. *Journal of Neurophysiology* 98 (6): 3360–3369.

52. Steyger, P.S., P.G. Gillespie, and R.A. Baird. 1998. Myosin Ibeta is located at tip link anchors in vestibular hair bundles. *The Journal of Neuroscience* 18 (12): 4603–4615.

53. Garcia, J.A., et al. 1998. Localization of myosin-Ibeta near both ends of tip links in frog sac-cular hair cells. *The Journal of Neuroscience* 18 (21): 8637–8647.

54. Schneider, M.E., et al. 2006. A new compartment at stereocilia tips defined by spatial and tem-poral patterns of myosin IIIa expression. *The Journal of Neuroscience* 26 (40): 10243–10252.

55. Gillespie, P.G., et al. 1999. Engineering of the myosin-ibeta nucleotide-binding pocket to cre-ate selective sensitivity to N(6)-modified ADP analogs. *The Journal of Biological Chemistry* 274 (44): 31373–31381.

56. Holt, J.R., et al. 2002. A chemical-genetic strategy implicates myosin-1c in adaptation by hair cells. *Cell* 108 (3): 371–381.

57. Stauffer, E.A., et al. 2005. Fast adaptation in vestibular hair cells requires myosin-1c activity. *Neuron* 47 (4): 541–553.

58. Geng, R., et al. 2013. Noddy, a mouse harboring a missense mutation in protocadherin-15, reveals the impact of disrupting a critical interaction site between tip-link cadherins in inner ear hair cells. *The Journal of Neuroscience* 33 (10): 4395–4404.

59. Araya-Secchi, R., B.L. Neel, and M. Sotomayor. 2016. An elastic element in the protocadherin-15 tip link of the inner ear. *Nature Communications* 7: 13458.
60. Powers, R.E., R. Gaudet, and M. Sotomayor. 2017. A partial calcium-free linker confers flexibility to inner-ear protocadherin-15. *Structure* 25: 482–495.
61. Reiners, J., et al. 2005. Photoreceptor expression of the Usher syndrome type 1 protein protocadherin 15 (USH1F) and its interaction with the scaffold protein harmonin (USH1C). *Molecular Vision* 11: 347–355.
62. Nie, H., et al. 2016. Plasma membrane targeting of protocadherin 15 is regulated by the Golgi-associated chaperone protein PIST. *Neural Plasticity* 2016: 8580675.
63. Pepermans, E., et al. 2014. The CD2 isoform of protocadherin-15 is an essential component of the tip-link complex in mature auditory hair cells. *EMBO Molecular Medicine* 6 (7): 984–992.
64. Webb, S.W., et al. 2011. Regulation of PCDH15 function in mechanosensory hair cells by alternative splicing of the cytoplasmic domain. *Development* 138 (8): 1607–1617.
65. Zhao, B., et al. 2014. TMIE is an essential component of the mechanotransduction machinery of cochlear hair cells. *Neuron* 84 (5): 954–967.
66. Senften, M., et al. 2006. Physical and functional interaction between protocadherin 15 and myosin VIIa in mechanosensory hair cells. *The Journal of Neuroscience* 26 (7): 2060–2071.
67. Ricci, A.J., A.C. Crawford, and R. Fettiplace. 2003. Tonotopic variation in the conductance of the hair cell mechanotransducer channel. *Neuron* 40 (5): 983–990.
68. Beurg, M., et al. 2009. Localization of inner hair cell mechanotransducer channels using high-speed calcium imaging. *Nature Neuroscience* 12 (5): 553–558.
69. Kurima, K., et al. 2002. Dominant and recessive deafness caused by mutations of a novel gene, TMC1, required for cochlear hair-cell function. *Nature Genetics* 30 (3): 277–284.
70. Keresztes, G., H. Mutai, and S. Heller. 2003. TMC and EVER genes belong to a larger novel family, the TMC gene family encoding transmembrane proteins. *BMC Genomics* 4 (1): 24.
71. Kurima, K., et al. 2003. Characterization of the transmembrane channel-like (TMC) gene family: functional clues from hearing loss and epidermodysplasia verruciformis. *Genomics* 82 (3): 300–308.
72. Kawashima, Y., et al. 2011. Mechanotransduction in mouse inner ear hair cells requires transmembrane channel-like genes. *The Journal of Clinical Investigation* 121 (12): 4796–4809.
73. Labay, V., et al. 2010. Topology of transmembrane channel-like gene 1 protein. *Biochemistry* 49 (39): 8592–8598.
74. Scheffer, D.I., et al. 2015. Gene expression by mouse inner ear hair cells during development. *The Journal of Neuroscience* 35 (16): 6366–6380.
75. Shen, J., et al. 2015. SHIELD: an integrative gene expression database for inner ear research. *Database: The Journal of Biological Databases and Curation* 2015: bav071.
76. Geleoc, G.S., and J.R. Holt. 2003. Developmental acquisition of sensory transduction in hair cells of the mouse inner ear. *Nature Neuroscience* 6 (10): 1019–1020.
77. Kurima, K., et al. 2015. TMC1 and TMC2 localize at the site of mechanotransduction in mammalian inner ear hair cell stereocilia. *Cell Reports* 12 (10): 1606–1617.
78. Maeda, R., et al. 2014. Tip-link protein protocadherin 15 interacts with transmembrane channel-like proteins TMC1 and TMC2. *Proceedings of the National Academy of Sciences of the United States of America* 111 (35): 12907–12912.
79. Beurg, M., et al. 2015. Subunit determination of the conductance of hair-cell mechanotransducer channels. *Proceedings of the National Academy of Sciences of the United States of America* 112 (5): 1589–1594.
80. Vreugde, S., et al. 2002. Beethoven, a mouse model for dominant, progressive hearing loss DFNA36. *Nature Genetics* 30 (3): 257–258.
81. Steel, K.P., and G.R. Bock. 1980. The nature of inherited deafness in deafness mice. *Nature* 288 (5787): 159–161.

82. Marcotti, W., et al. 2006. Tmc1 is necessary for normal functional maturation and survival of inner and outer hair cells in the mouse cochlea. *The Journal of Physiology* 574 (Pt 3): 677–698.

83. Kim, K.X., et al. 2013. The role of transmembrane channel-like proteins in the operation of hair cell mechanotransducer channels. *The Journal of General Physiology* 142 (5): 493–505.

84. Pan, B., et al. 2013. TMC1 and TMC2 are components of the mechanotransduction channel in hair cells of the mammalian inner ear. *Neuron* 79 (3): 504–515.

85. Cunningham, C.L., et al. 2017. The murine catecholamine methyltransferase mTOMT is essential for mechanotransduction by cochlear hair cells. *eLife* 6: e24318.

86. Erickson, T., et al. 2017. Integration of Tmc1/2 into the mechanotransduction complex in zebrafish hair cells is regulated by transmembrane O-methyltransferase (Tomt). *eLife* 6: e28474.

87. Ahmed, Z.M., et al. 2008. Mutations of LRTOMT, a fusion gene with alternative reading frames, cause nonsyndromic deafness in humans. *Nature Genetics* 40 (11): 1335–1340.

88. Du, X., et al. 2008. A catechol-O-methyltransferase that is essential for auditory function in mice and humans. *Proceedings of the National Academy of Sciences of the United States of America* 105 (38): 14609–14614.

89. Fettiplace, R. 2016. Is TMC1 the hair cell mechanotransducer channel? *Biophysical Journal* 111 (1): 3–9.

90. Corey, D.P., and J.R. Holt. 2016. Are TMCs the mechanotransduction channels of vertebrate hair cells? *The Journal of Neuroscience* 36 (43): 10921–10926.

91. Wu, Z., and U. Muller. 2016. Molecular identity of the mechanotransduction channel in hair cells: not quiet there yet. *The Journal of Neuroscience* 36 (43): 10927–10934.

92. Chatzigeorgiou, M., et al. 2013. *tmc-1* encodes a sodium-sensitive channel required for salt chemosensation in C. elegans. *Nature* 494 (7435): 95–99.

93. Longo-Guess, C.M., et al. 2007. Targeted knockout and *lacZ* reporter expression of the mouse *Tmhs* deafness gene and characterization of the *hscy-2J* mutation. *Mammalian Genome* 18 (9): 646–656.

94. Xiong, W., et al. 2012. TMHS is an integral component of the mechanotransduction machinery of cochlear hair cells. *Cell* 151 (6): 1283–1295.

95. Petit, M.M., et al. 1999. LHFP, a novel translocation partner gene of HMGIC in a lipoma, is a member of a new family of LHFP-like genes. *Genomics* 57 (3): 438–441.

96. Longo-Guess, C.M., et al. 2005. A missense mutation in the previously undescribed gene Tmhs underlies deafness in hurry-scurry (hscy) mice. *Proceedings of the National Academy of Sciences of the United States of America* 102 (22): 7894–7899.

97. Gunzel, D. 2017. Claudins: vital partners in transcellular and paracellular transport coupling. *Pflügers Archiv* 469 (1): 35–44.

98. Kar, R., et al. 2012. Biological role of connexin intercellular channels and hemichannels. *Archives of Biochemistry and Biophysics* 524 (1): 2–15.

99. Syeda, R., et al. 2016. LRRC8 proteins form volume-regulated anion channels that sense ionic strength. *Cell* 164 (3): 499–511.

100. Jackson, A.C., and R.A. Nicoll. 2011. The expanding social network of ionotropic glutamate receptors: TARPs and other transmembrane auxiliary subunits. *Neuron* 70 (2): 178–199.

101. Kalay, E., et al. 2006. Mutations in the lipoma HMGIC fusion partner-like 5 (LHFPL5) gene cause autosomal recessive nonsyndromic hearing loss. *Human Mutation* 27 (7): 633–639.

102. Shabbir, M.I., et al. 2006. Mutations of human TMHS cause recessively inherited non-syndromic hearing loss. *Journal of Medical Genetics* 43 (8): 634–640.

103. Mitchem, K.L., et al. 2002. Mutation of the novel gene Tmie results in sensory cell defects in the inner ear of spinner, a mouse model of human hearing loss DFNB6. *Human Molecular Genetics* 11 (16): 1887–1898.

104. Karuppasamy, S., et al. 2011. Subcellular localization of the transmembrane inner ear (Tmie) protein in a stable Tmie-expressing cell line. *Laboratory Animal Research* 27 (4): 339–342.

105. Su, M.C., et al. 2008. Expression and localization of Tmie in adult rat cochlea. *Histochemistry and Cell Biology* 130 (1): 119–126.
106. O'Hagan, R., M. Chalfie, and M.B. Goodman. 2005. The MEC-4 DEG/ENaC channel of Caenorhabditis elegans touch receptor neurons transduces mechanical signals. *Nature Neuroscience* 8 (1): 43–50.
107. Naz, S., et al. 2002. Mutations in a novel gene, TMIE, are associated with hearing loss linked to the DFNB6 locus. *American Journal of Human Genetics* 71 (3): 632–636.
108. Cho, K.I., et al. 2006. The circling mouse (C57BL/6J-cir) has a 40-kilobase genomic deletion that includes the transmembrane inner ear (tmie) gene. *Comparative Medicine* 56 (6): 476–481.
109. Chung, W.H., et al. 2007. Cochlear pathology of the circling mouse: a new mouse model of DFNB6. *Acta Oto-Laryngologica* 127 (3): 244–251.
110. Seki, N., et al. 1999. Structure, expression profile and chromosomal location of an isolog of DNA-PKcs interacting protein (KIP) gene. *Biochimica et Biophysica Acta* 1444 (1): 143–147.
111. Riazuddin, S., et al. 2012. Alterations of the CIB2 calcium- and integrin-binding protein cause Usher syndrome type 1J and nonsyndromic deafness DFNB48. *Nature Genetics* 44 (11): 1265–1271.
112. Hager, M., et al. 2008. Cib2 binds integrin alpha7Bbeta1D and is reduced in laminin alpha2 chain-deficient muscular dystrophy. *The Journal of Biological Chemistry* 283 (36): 24760–24769.
113. Giese, A.P.J., et al. 2017. CIB2 interacts with TMC1 and TMC2 and is essential for mechanotransduction in auditory hair cells. *Nature Communications* 8 (1): 43.
114. Gentry, H.R., et al. 2005. Structural and biochemical characterization of CIB1 delineates a new family of EF-hand-containing proteins. *The Journal of Biological Chemistry* 280 (9): 8407–8415.
115. Blazejczyk, M., et al. 2009. Biochemical characterization and expression analysis of a novel EF-hand Ca2+ binding protein calmyrin2 (Cib2) in brain indicates its function in NMDA receptor mediated Ca2+ signaling. *Archives of Biochemistry and Biophysics* 487 (1): 66–78.
116. Huang, H., J.N. Bogstie, and H.J. Vogel. 2012. Biophysical and structural studies of the human calcium- and integrin-binding protein family: understanding their functional similarities and differences. *Biochemistry and Cell Biology* 90 (5): 646–656.
117. Zhu, W., et al. 2017. CIB2 negatively regulates oncogenic signaling in ovarian cancer via sphingosine kinase 1. *Cancer Research* 77 (18): 4823–4834.
118. Rato, S., et al. 2010. Novel HIV-1 knockdown targets identified by an enriched kinases/phosphatases shRNA library using a long-term iterative screen in Jurkat T-cells. *PLoS One* 5 (2): e9276.
119. Godinho-Santos, A., et al. 2016. CIB1 and CIB2 are HIV-1 helper factors involved in viral entry. *Scientific Reports* 6: 30927.
120. Patel, K., et al. 2015. A novel C-terminal CIB2 (Calcium and Integrin Binding Protein 2) mutation associated with non-syndromic hearing loss in a hispanic family. *PLoS One* 10 (10): e0133082.
121. Seco, C.Z., et al. 2016. Novel and recurrent CIB2 variants, associated with nonsyndromic deafness, do not affect calcium buffering and localization in hair cells. *European Journal of Human Genetics* 24 (4): 542–549.
122. Michel, V., et al. 2017. CIB2, defective in isolated deafness, is key for auditory hair cell mechanotransduction and survival. *EMBO Molecular Medicine* 9 (12): 1711–1731.
123. Booth, K.T., et al. 2018. Variants in CIB2 cause DFNB48 and not USH1J. *Clinical Genetics*. https://doi.org/10.1111/cge.13170.
124. Zou, J., et al. 2017. The roles of USH1 proteins and PDZ domain-containing USH proteins in USH2 complex integrity in cochlear hair cells. *Human Molecular Genetics* 26 (3): 624–636.
125. Wang, Y., et al. 2017. Loss of CIB2 causes profound hearing loss and abolishes mechanoelectrical transduction in mice. *Frontiers in Molecular Neuroscience* 10: 401.

126. Shotwell, S.L., R. Jacobs, and A.J. Hudspeth. 1981. Directional sensitivity of individual vertebrate hair cells to controlled deflection of their hair bundles. *Annals of the New York Academy of Sciences* 374: 1–10.
127. Kindt, K.S., G. Finch, and T. Nicolson. 2012. Kinocilia mediate mechanosensitivity in developing zebrafish hair cells. *Developmental Cell* 23 (2): 329–341.
128. Marcotti, W., et al. 2014. Transduction without tip links in cochlear hair cells is mediated by ion channels with permeation properties distinct from those of the mechano-electrical transducer channel. *The Journal of Neuroscience* 34 (16): 5505–5514.
129. Michalski, N., et al. 2007. Molecular characterization of the ankle-link complex in cochlear hair cells and its role in the hair bundle functioning. *The Journal of Neuroscience* 27 (24): 6478–6488.
130. Stepanyan, R., and G.I. Frolenkov. 2009. Fast adaptation and Ca2+ sensitivity of the mechanotransducer require myosin-XVa in inner but not outer cochlear hair cells. *The Journal of Neuroscience* 29 (13): 4023–4034.
131. Beurg, M., K.X. Kim, and R. Fettiplace. 2014. Conductance and block of hair-cell mechanotransducer channels in transmembrane channel-like protein mutants. *The Journal of General Physiology* 144 (1): 55–69.
132. Beurg, M., et al. 2016. Development and localization of reverse-polarity mechanotransducer channels in cochlear hair cells. *Proceedings of the National Academy of Sciences of the United States of America* 113 (24): 6767–6772.
133. Wu, Z., et al. 2017. Mechanosensory hair cells express two molecularly distinct mechanotransduction channels. *Nature Neuroscience* 20 (1): 24–33.
134. Coste, B., et al. 2010. Piezo1 and Piezo2 are essential components of distinct mechanically activated cation channels. *Science* 330 (6000): 55–60.
135. Woo, S.H., et al. 2014. Piezo2 is required for Merkel-cell mechanotransduction. *Nature* 509 (7502): 622–626.
136. Ranade, S.S., et al. 2014. Piezo2 is the major transducer of mechanical forces for touch sensation in mice. *Nature* 516 (7529): 121–125.
137. Woo, S.H., et al. 2015. Piezo2 is the principal mechanotransduction channel for proprioception. *Nature Neuroscience* 18 (12): 1756–1762.
138. Nonomura, K., et al. 2017. Piezo2 senses airway stretch and mediates lung inflation-induced apnoea. *Nature* 541 (7636): 176–181.
139. Coste, B., et al. 2013. Gain-of-function mutations in the mechanically activated ion channel PIEZO2 cause a subtype of Distal Arthrogryposis. *Proceedings of the National Academy of Sciences of the United States of America* 110 (12): 4667–4672.
140. McMillin, M.J., et al. 2014. Mutations in PIEZO2 cause Gordon syndrome, Marden-Walker syndrome, and distal arthrogryposis type 5. *American Journal of Human Genetics* 94 (5): 734–744.
141. Delle Vedove, A., et al. 2016. Biallelic loss of proprioception-related PIEZO2 causes muscular atrophy with perinatal respiratory distress, arthrogryposis, and scoliosis. *American Journal of Human Genetics* 99 (5): 1206–1216.
142. Chesler, A.T., et al. 2016. The role of PIEZO2 in human mechanosensation. *The New England Journal of Medicine* 375 (14): 1355–1364.
143. Indzhykulian, A.A., et al. 2013. Molecular remodeling of tip links underlies mechanosensory regeneration in auditory hair cells. *PLoS Biology* 11 (6): e1001583.

Chapter 5
Mechanotransduction and Inner Ear Function

Wei Xiong

Abstract Hair-cell mechanotransduction (MET) plays so important a role in auditory sensation that its malfunction introduces severe hearing impairment. In this chapter, we listed all of the known genes linked to MET functionality, which was collectively learned from studies on mouse models and human genetics. It might give us implication in how the hearing disorder happens due to MET defect. Based on this knowledge, we would like to discuss some physiological significance of MET and potential therapeutics to treat hearing loss.

Keywords Human genetics · Mouse models · Cochlear tuning · Inner ear dysfunction · Gene therapy

5.1 Implication in Mechanotransduction from Human Genetics and Mouse Models

So far, 13 genes have been characterized as MET genes that encode proteins directly contributing MET structurally and/or functionally. These proteins are components of the tip link or of complex at tip-link ends. As introduced in the previous chapters, CDH23 is the upper part of tip link and PCDH15 is the lower part of tip link. In upper tip-link density (UTLD), harmonin/USH1C, Myo7a, and Sans interact with intracellular domains of CDH23 that regulate the tension of tip link by anchoring CDH23 with actin-based cytoskeleton. Lhfpl5/TMHS, TMC1, TMC2, and TMIE form the MET channel complex that associate the membrane insertion site of PCDH15 in lower tip-link density (LTLD). CIB2 and TOMT are soluble proteins, but they interact with TMC1 (likely TMC2 too) and destroyed MET response when perturbed. Interestingly, TOMT is not physically distributed in stereocilia, which

W. Xiong (✉)
School of Life Sciences, IDG/McGovern Institute for Brain Research, Tsinghua University, Beijing, China
e-mail: wei_xiong@mail.tsinghua.edu.cn

© The Author(s) 2018
W. Xiong, Z. Xu (eds.), *Mechanotransduction of the Hair Cell*,
SpringerBriefs in Biochemistry and Molecular Biology,
https://doi.org/10.1007/978-981-10-8557-4_5

indicates TOMT assembles and transports MET components into stereocilia. Myo15a and whirlin are at the top of stereocilia and modulate organization of staircase shape of hair bundle.

Study on Usher syndrome has brought us a better understanding on MET mechanisms. It was indicated that the deaf phenotype of Usher syndrome patients somehow is a direct manifestation of transduction problem. Currently, USH genes include MYO7A (USH1B), USH1C/harmonin (USH1C), CDH23 (USH1D), PCDH15 (USH1F), USH1G (USH1G), usherin (USH2A), ADGRV1 (USH2C), whirlin (USH2D), and CLRN1 (USH3A). CIB2 was just reported in 2012 as a candidate gene of USH1J but was recently ruled out for no blind phenotype confirmed [1–3]. When studying the MET components and function, human genetics and mouse models have contributed significantly. Here I would like to take harmonin (USH1C) and PCDH15 (USH1F) for examples.

Two groups identified *USH1C/harmonin* from the USH1C loci by human genetics study [4, 5]. After analysing the affected individuals from three consanguineous families with hyperinsulinism, profound congenital sensorineural deafness, enteropathy, and renal tubular dysfunction, Bitner-Glindzicz et al. linked the causative gene to *harmonin* and ruled out the possible cause by mutation in ABCC8 or KCNJ11 from a 122-kb deletion of chromosome 11 short arm [4]. After analysing three consanguineous families, two families were due to indel-induced early stop codon, and one family resulted from an expansion of variable number of tandem repeat (VNTR) of a 45-bp element in intron 5 in *harmonin* gene [5]. Harmonin has three isoforms, namely, a, b, and c. By immunofluorescence, harmonin is enriched in hair cells [5]. Harmonin b is the longest version and the main isoform for hair-cell function that consists of three PDZ domains, one coiled-coil (CC) domain, and a proline-serine-threonine (PST)-rich domain [6, 7]. Harmonin b facilities actin filaments assemble into bundles. Biochemically, harmonin and cadherin 23 bind to each other to form a complex. It is assumed that harmonin b likely anchors CDH23 to the actin cytoskeleton as an anchorage. Besides, harmonin b interacts directly with myosin VIIa [6, 7]. Later, evidence showed that oligomerization of harmonin b further stabilizes the UTLD by interacting with other proteins such as Myo7a and Sans [8]. Harmonin was named from the Greek word *harmonia* that means *assembling* [5], which was quite precise. In 2003, two spontaneous mutant USH1C mice lines were reported that carried two new recessive mutations causing circling behaviour and deafness. One was named *deaf circler* (*dfcr*) and the other was *deaf circler 2 Jackson* (*dfcr-2J*). The *dfcr* mutation possesses a 12.8-kb intragenic deletion that excludes exons encoding CC1, CC2, and PST. There is a 1-bp deletion in *dfcr-2J* that generates a premature stop codon by introducing a frameshift that eliminates domains from CC2 [9]. Later an *USH1C* knockout mouse line and a *PDZ2^{AAA}* knockin mouse line with disrupted harmonin-CDH23 interaction were generated [10, 11]. Take the advantage of multiple mutant lines including *dfcr*, *dfcr-2j*, *PDZ2^{AAA}* knockin, and *USH1C* knockout, the tip link and transducer properties were studied to understand how harmonin plays roles in MET. The immunohistological data showed that harmonin is a component at UTLDs in hair cells of adult mice. Harmonin PDZ2 domains mediate CDH23 interaction that controls harmonin localization in stereocilia. Domains of CC1, CC2, and PST contribute F-actin inter-action [10–12]. Structural study further provided insight into molecular interaction of the harmonin PDZ2 and the CDH23 carboxyl tail in a multidentate binding mode [13]. In *USH1C* knockout, the MET currents were drastically reduced. In *dfcr* mice missing CC1, CC2, and PST domains, the truncated harmonin moves to the top of the stereocilia, and the MET sensitivity of hair bundles is reduced but with the MET intensity reserved. In *dfcr-2J* OHCs, the transduction current amplitude and sensi-tivity are both decreased due to loss of CC2, PST, and PDZ3 [11, 12]. All the study based on the four lines shed light on the working mechanisms of harmonin that PDZ2, CC, and PST domains are essential for CDH23 and actin cytoskeleton anchorage at UTLD. Harmonin as a scaffold protein regulates the tension of tip link and MET sensitivity.

PCDH15 gene was identified from a mouse line called *Ames waltzer* (*av*) [14] with deafness, imbalance, and degeneration of inner ear neuroepithelia that was characterized 30 years ago [15]. From the study of human families, PCDH15 was confirmed in *USH1F* loci harboured in human chromosome 10q21–22 [16–18]. Immunofluorescence study showed that PCDH15 is highly expressed along the length of stereocilia [19]. In 2006, Ahmed et al. reported that four major protocad-herin-15 transcripts (CD1, CD2, CD3, and a secreted version) were expressed in the hair cells of mouse inner ear. The isoforms had distinct spatiotemporal expression patterns in developing and mature hair cells [20]. However, they could not firmly judge which type of linkages PCDH15 belonged to. In 2007, Kazmierczak et al. showed that PCDH15 is actually the lower part of the tip link with CDH23 as the upper part [21]. Taking advantage of the stunning antibody work, they calculated that the 180 nm length of tip link is composed of 50 nm PCDH15 and 130 nm CDH23. It raised again the question what is the function for different CD (cytosol domain) and CR (common region) of PCDH15. Later, mouse lines with isoform-specific knockout were generated to study function of each CD isoform. However, lack of either CD did not affect too much of the MET, but only CD2 knockout dis-turbed the polarity of the hair bundle due to kinocilium-stereocilium link defect [22]. In another study, CD2 domain was specifically deleted in adult hair cells, which led to almost complete loss of tip links in mature hair cells, suggesting that PCDH15-CD2 is essential for tip-link formation [71]. Further, PCDH15 was found to have direct interaction with other MET components including LHFPL5, TMC1, TMC2, TMIE, and TOMT [23–26]. In zebrafish, hair-cell-enriched paralog Pcdh15a also has two isoforms Pcdh15a-CD1 and Pcdh15a-CD3. Interestingly, CD regions are not required for Pcdh15a localization, and Pcdh15a-CR was sufficient for MET complex formation and function [27].

Harmonin and PCDH15 are good cases to show how delicate a protein functions in tissue-, isoform-, and domain-specific patterns. With different mutations, the pro-tein product may play a different role in cellular function. This might explain why human patients suffer different disorders even if they encounter the defect in the same gene. Early study from human genetics has greatly guided the progress on hearing research. However, due to modernized life style, consanguineous families have been less and less. Now to understand the biological and pathological mechanisms, animal models are precious for scientific and clinic study as a significant resource for study on molecular function, cellular function, and physiological function of the inner ear. Since it is still difficult to obtain a whole picture of MET working mechanisms based on currently known components, it is predictable there will be more MET components to be characterized especially from mouse genetics.

5.2 Frequency Dependency of Mechanotransduction and Cochlear Tuning

The inner ear is a natural digitizer by which human can decode the intensity and frequency information of sound. It has been known that basilar membrane is the key apparatus that oscillates to certain frequencies at assigned locations in a nonlinear way [28–32]. By this way, basilar membrane and motorized OHCs enhance the vibration of each other reciprocally, which was identified with prestin as the molecular motor [33, 34]. The frequency is selectively conveyed to the best-responded IHCs. Biophysical data on cochlear tuning has been collected from different species but with guinea pigs studied most thoroughly. With frequency sweep method, Evans estimated the frequency-threshold curve of a single cochlear fibre in guinea pigs [35]. Actually the sharp tuning curves were observed in guinea pigs at several aspects including basilar membrane [31], IHCs [36, 37], and auditory fibres [35]. The receptor potential of IHCs consisted of an AC response following the tone burst and a DC response enveloping the tone burst [36, 37]. For thorough description, one can refer reviews from others [38, 39].

In lower vertebrate such as turtle and chicken, this frequency selectivity was not the same case. It has been found that an electrical resonance happened during current injection in auditory hair cells [40]. This tuning range is consistent with the frequencies that the hair cells could represent. For example, hair cells of the turtles possessed frequencies from several dozens to about 700 Hz, while hair cells of the bullfrogs were from 80 to 160 Hz [41, 42]. There is a strong correlation between acoustic stimulation frequencies and electrical characteristic frequencies in recorded hair cells. Moreover, there is a nonlinear relationship between the sound intensities [40, 43], which was also confirmed in solitary hair cells of the bullfrogs by whole-cell current-clamp mode. Pharmacological study has shown that voltage-gated calcium channels and calcium-activated potassium (BK) channels were contributed to this electrical tuning in turtles and bullfrogs [41, 42, 44]. Consistently, the alternative splicing and channel activity of BK channels were graded along the cochlear coils [45, 46]. However, cochlear tuning largely relied on prestin-based cellular motility instead of BK channel fluctuation in mammalian cochleae [47, 48].

Interestingly, certain hair cells have graded morphological and biophysical properties to match their best frequencies. In mammalian OHCs, length of somas and bundles becomes longer from high-frequency cells to low-frequency cells.

Nevertheless, the MET current amplitude and unitary MET channel conductance are smaller from high-frequency OHCs to low-frequency OHCs, which was examined by whole-cell recording in turtle [41, 49], in gerbil [50], and in rats [51]. In addition, the adaptation also alters from low-frequency cells to high-frequency cells [52]. It is worth noting that the tonotopy is also reflected by single-channel conductance variance [49, 51]. Since MET channel is composed of multiple components, questions are raised whether any molecules contribute to this tonotopy and whether the micro-environment of cochlea mediates the gradient of channel properties. It has been shown that the tonotopic distribution of MET current was gone when deleting LHFPL5 and TMC1 [53–55], suggesting the composition of MET channels may adapt to tonotopy. However, omitting other MET components directly destroys the MET currents except LHFPL5 and TMCs, which leads this question to be further considered. Other puzzles include whether developmental factors and physical factor such as calcium and endocochlear potential play roles to modulate the channel tonotopy.

5.3 Transduction Defect and Inner Ear Dysfunction

Undoubtedly, transduction efficiency is directly linked to the inner ear function and to the auditory perception. As we discussed in the previous chapters, anything that happens to the transduction machine, either structurally or functionally, all causes hearing loss. The MET complex is so well-designed machinery that knockout of a single part in the complex triggers the deafness phenotype. So far, there are 134 genetic loci linked to nonsyndromic deafness and 100 genes identified according to *The Hereditary Hearing loss Homepage* (http://hereditaryhearingloss.org data updated in Oct. 2017). These genes are mainly functioning in hair cells, stria vascularis, ion homeostasis, and inner ear development that are directly or indirectly contributing to the transduction. Among those genes, 32 genes are hair-cell-specific genes, and around one third are nonsyndromic deafness genes.

As a most common sensory disease, hearing loss and deafness are affecting millions of people all over the world. Especially, single mutation of susceptive genes for hearing directly causes congenital deafness that largely deteriorates life quality of patients. It implies how fragile the human gene is and the door to hearing health is easily closed. Of course, the inner ear is such a unique structure, and the 100 inner ear specific genes were organized to contribute the structure and function, implying that each gene was indispensable and seriously collected along the evolution. Fortunately, the anatomical, developmental, and genetic mechanisms of the inner ear were relatively well understood, which provides a window to probably prevent and treat the hearing disorders. Comparing to central neurological disease, it is easier to manipulate accurately the spatiotemporal pattern of a protein expression by in vivo genetic correction. More than two decades ago, people have tried intensively to obtain a consistent gene delivery approach ranging from virus to electroporation

[56–58]. The first proof of concept was from the in vivo injection of vesicular glutamate transporter-3 (VGLUT3) gene into the VGLUT3 knockout mice [59]. It provided a successful case that viral delivery of VGLUT3 into P10–12 knockout mice rescued hearing significantly after 3 weeks. Evaluated by audiometry, the hearing was back to normal in 7–14 days and sustained for at least 7 weeks for most of the treated mice even for 1.5 years for two mice. If delivered in postnatal 1–3 days, the transfection was more robust, and the hearing restore was longer. However, the application was still limited because the AAV1 infected IHCs mostly but not OHCs and it was a transient expression with short-term rescue effect. Meanwhile, several lines of progress were approached successfully by other therapies such as stem cell therapy, pharmacological therapy, and antisense oligo therapy [60–62]. And achievements were also obtained for vision restoration [63, 64]. These progresses further improved the confidence of the researchers to get a practical therapy in the near future. Gene therapy maybe prioritized to other biological therapeutics since the recently developed genome editing technique such as CRISPR/Cas9 opened a new route to make accurate alteration at single nucleotide level. We even could imagine a scenario that a causative gene mutation is detected by a prenatal diagnosis and then corrected by a postnatal genome manipulation in vitro or in vivo. There has been a successful application of genome editing on Duchenne muscular dystrophy in *mdx* mouse models [65–67]. The mutation in *mdx* exon 23 caused an early stop codon, but it was predicted that the protein was relatively normal with deletion of exon 23. The strategy also took the advantage that the skeletal muscle with a mosaic pattern of corrected myofibres showed rescued adequate dystrophin expression. The similar strategy could be applied on hearing restore in the inner ear. However, the first line was to get a reliable gene delivery plan to infect both OHCs and IHCs with enough cell numbers. Excitingly, a new gene delivery approach based on cationic lipid-mediated protein delivery was developed with promising effect. In mouse inner ear, introduction of Cre recombinase and Cas9-sgRNA complexes in vivo showed effective gene knockout 13–20% of OHCs [68]. More recently, a newly synthesized AAV (Anc80L65) was proved to infect 50% IHCs and 35% OHCs by average when injected with 1.36×10^{12} titres [69]. In a side-by-side paper, it was shown that wild-type harmonin via Anc80L65 transduced 80–90% of sensory hair cells in *USH1C c.216AA* mice, and the recovery was found from hair bundle morphology to MET response, to ABR (auditory brainstem response) and DPOAE (distorted product optoacoustic emission), and to auditory and balance behaviour. Nevertheless, the most exciting progress is the genome editing application on genetically induced hearing loss. With cationic lipid-mediated Cas9::sgRNA complex delivery in vivo previously developed by the same laboratory, they showed that allele-specific disruption of disease gene relieved deaf phenotype in *Beethoven* mice that carried a dominant mutant of *tmc1* gene [70]. The first AAV-based gene therapy for inherited disease was just approved by FDA in Dec. 2017, which had a huge social impact. The future will be bright for in vivo gene modification-based therapy that will firmly step in public health system.

5.4   Discussion

So far, it has been well appreciated that MET is orchestrated by a complex including more than a dozen proteins from tip-link component to pore-forming subunit. As for the reason why so many components cooperate to perform auditory transduction, we do not quite know yet. Further, why it is so complicated for the gating mechanism is also not quite understood due to lack of clear picture of the channel identity. Luckily, the massive studies from human genetics and mouse models have contributed a lot to fill in our knowledge gap. Especially the genetic data provided the functional evidence with a significant note about the mechanisms underlying the normal hearing and deafness. At least two aspects are emerging to be solved: cochlear tuning and gene therapy of hearing. As MET contributes in the cochlear transduction deeply more than we have thought, it might not just simply convert the sound-induced vibration but also shape its response to fit frequency specificity. It is wondered whether there is molecular basis to functionally modify the channel composition or it is just a developmental manifestation. Last, it is exciting to have much progress for gene therapy application on genetically induced hearing loss in recent years. It shed light on the deaf patients who are still struggling with this nonlethal but painful disease.

References

1. Michel, V., et al. 2017. CIB2, defective in isolated deafness, is key for auditory hair cell mechanotransduction and survival. *EMBO Molecular Medicine* 9: 1711–1731.
2. Riazuddin, S., et al. 2012. Alterations of the CIB2 calcium- and integrin-binding protein cause Usher syndrome type 1J and nonsyndromic deafness DFNB48. *Nature Genetics* 44 (11): 1265–1271.
3. Booth, K.T., et al. 2017. Variants in *CIB2* cause DFNB48 and not USH1J. *Clinical Genetics*. https://doi.org/10.1111/cge.13170.
4. Bitner-Glindzicz, M., et al. 2000. A recessive contiguous gene deletion causing infantile hyperinsulinism, enteropathy and deafness identifies the Usher type 1C gene. *Nature Genetics* 26 (1): 56–60.
5. Verpy, E., et al. 2000. A defect in harmonin, a PDZ domain-containing protein expressed in the inner ear sensory hair cells, underlies Usher syndrome type 1C. *Nature Genetics* 26 (1): 51–55.
6. Siemens, J., et al. 2002. The Usher syndrome proteins cadherin 23 and harmonin form a complex by means of PDZ-domain interactions. *Proceedings of the National Academy of Sciences of the United States of America* 99 (23): 14946–14951.
7. Boeda, B., et al. 2002. Myosin VIIa, harmonin and cadherin 23, three Usher I gene products that cooperate to shape the sensory hair cell bundle. *EMBO Journal* 21 (24): 6689–6699.
8. Weil, D., et al. 2003. Usher syndrome type I G (USH1G) is caused by mutations in the gene encoding SANS, a protein that associates with the USH1C protein, harmonin. *Human Molecular Genetics* 12 (5): 463–471.
9. Johnson, K.R., et al. 2003. Mouse models of USH1C and DFNB18: phenotypic and molecular analyses of two new spontaneous mutations of the Ush1c gene. *Human Molecular Genetics* 12 (23): 3075–3086.

10. Lefevre, G., et al. 2008. A core cochlear phenotype in USH1 mouse mutants implicates fibrous links of the hair bundle in its cohesion, orientation and differential growth. *Development* 135 (8): 1427–1437.
11. Grillet, N., et al. 2009. Harmonin mutations cause mechanotransduction defects in cochlear hair cells. *Neuron* 62 (3): 375–387.
12. Michalski, N., et al. 2009. Harmonin-b, an actin-binding scaffold protein, is involved in the adaptation of mechanoelectrical transduction by sensory hair cells. *Pflügers Archiv – European Journal of Physiology* 459 (1): 115–130.
13. Pan, L.F., et al. 2009. Assembling stable hair cell tip link complex via multidentate interactions between harmonin and cadherin 23. *Proceedings of the National Academy of Sciences of the United States of America* 106 (14): 5575–5580.
14. Alagramam, K.N., et al. 2001. The mouse Ames waltzer hearing-loss mutant is caused by mutation of Pcdh15, a novel protocadherin gene. *Nature Genetics* 27 (1): 99–102.
15. Osako, S., and D.A. Hilding. 1971. Electron microscopic studies of capillary permeability in normal and Ames Waltzer deaf mice. *Acta Oto-Laryngologica* 71 (5): 365–376.
16. Ahmed, Z.M., et al. 2001. Mutations of the protocadherin gene PCDH15 cause Usher syndrome type 1F. *American Journal of Human Genetics* 69 (1): 25–34.
17. Alagramam, K.N., et al. 2001. Mutations in the novel protocadherin PCDH15 cause Usher syndrome type 1F. *Human Molecular Genetics* 10 (16): 1709–1718.
18. Ben-Yosef, T., et al. 2003. Brief report – a mutation of PCDH15 among Ashkenazi Jews with the type 1 Usher syndrome. *New England Journal of Medicine* 348 (17): 1664–1670.
19. Ahmed, Z.M., et al. 2003. PCDH15 is expressed in the neurosensory epithelium of the eye and ear and mutant alleles are responsible for both USH1F and DFNB23. *Human Molecular Genetics* 12 (24): 3215–3223.
20. ———. 2006. The tip-link antigen, a protein associated with the transduction complex of sensory hair cells, is protocadherin-15. *The Journal of Neuroscience* 26 (26): 7022–7034.
21. Kazmierczak, P., et al. 2007. Cadherin 23 and protocadherin 15 interact to form tip-link filaments in sensory hair cells. *Nature* 449 (7158): 87–91.
22. Webb, S.W., et al. 2011. Regulation of PCDH15 function in mechanosensory hair cells by alternative splicing of the cytoplasmic domain. *Development* 138 (8): 1607–1617.
23. Xiong, W., et al. 2012. TMHS is an integral component of the mechanotransduction machinery of cochlear hair cells. *Cell* 151 (6): 1283–1295.
24. Zhao, B., et al. 2014. TMIE is an essential component of the mechanotransduction machinery of cochlear hair cells. *Neuron* 84 (5): 954–967.
25. Cunningham, C.L., et al. 2017. The murine catecholamine methyltransferase mTOMT is essential for mechanotransduction by cochlear hair cells. *Elife* 6: e24318.
26. Maeda, R., et al. 2014. Tip-link protein protocadherin 15 interacts with transmembrane channel-like proteins TMC1 and TMC2. *Proceedings of the National Academy of Sciences of the United States of America* 111 (35): 12907–12912.
27. ———. 2017. Functional analysis of the transmembrane and cytoplasmic domains of Pcdh15a in zebrafish hair cells. *The Journal of Neuroscience* 37 (12): 3231–3245.
28. Von Bekesy, G. 1947. The variation of phase along the basilar membrane with sinusoidal vibrations. *Journal of the Acoustical Society of America* 19 (3): 452–460.
29. Gold, T. 1948. Hearing. 2. The physical basis of the action of the cochlea. *Proceedings of the Royal Society Series B-Biological Sciences* 135 (881): 492–498.
30. Gold, T., and R.J. Pumphrey. 1948. Hearing. 1. The cochlea as a frequency analyzer. *Proceedings of the Royal Society Series B-Biological Sciences* 135 (881): 462–491.
31. Rhode, W.S. 1971. Observations of vibration of basilar membrane in squirrel monkeys using Mossbauer technique. *Journal of the Acoustical Society of America* 49 (4): 1218.
32. Kemp, D.T. 1978. Stimulated acoustic emissions from within human auditory-system. *Journal of the Acoustical Society of America* 64 (5): 1386–1391.
33. Brownell, W.E., et al. 1985. Evoked mechanical responses of isolated cochlear outer hair cells. *Science* 227 (4683): 194–196.

34. Zheng, J., et al. 2000. Prestin is the motor protein of cochlear outer hair cells. *Nature* 405 (6783): 149–155.
35. Evans, E.F. 1972. The frequency response and other properties of single fibres in the guinea-pig cochlear nerve. *The Journal of Physiology* 226 (1): 263–287.
36. Russell, I.J., and P.M. Sellick. 1977. Tuning properties of cochlear hair cells. *Nature* 267 (5614): 858–860.
37. ———. 1978. Intracellular studies of hair cells in the mammalian cochlea. *The Journal of Physiology* 284: 261–290.
38. Dallos, P. 1992. The active cochlea. *The Journal of Neuroscience* 12 (12): 4575–4585.
39. Robles, L., and M.A. Ruggero. 2001. Mechanics of the mammalian cochlea. *Physiological Reviews* 81 (3): 1305–1352.
40. Crawford, A.C., and R. Fettiplace. 1981. An electrical tuning mechanism in turtle cochlear hair cells. *The Journal of Physiology* 312: 377–412.
41. Art, J.J., and R. Fettiplace. 1987. Variation of Membrane-Properties in Hair-Cells Isolated from the Turtle Cochlea. *The Journal of Physiology* 385: 207–242.
42. Hudspeth, A.J., and R.S. Lewis. 1988. A model for electrical resonance and frequency tuning in saccular hair cells of the bull-frog, Rana catesbeiana. *The Journal of Physiology* 400: 275–297.
43. Crawford, A.C., and R. Fettiplace. 1980. The frequency selectivity of auditory nerve fibres and hair cells in the cochlea of the turtle. *The Journal of Physiology* 306: 79–125.
44. Tucker, T.R., and R. Fettiplace. 1996. Monitoring calcium in turtle hair cells with a calcium-activated potassium channel. *The Journal of Physiology* 494 (Pt 3): 613–626.
45. Navaratnam, D.S., et al. 1997. Differential distribution of $Ca2+$-activated $K+$ channel splice variants among hair cells along the tonotopic axis of the chick cochlea. *Neuron* 19 (5): 1077–1085.
46. Ramanathan, K., et al. 1999. A molecular mechanism for electrical tuning of cochlear hair cells. *Science* 283 (5399): 215–217.
47. Oliver, D., et al. 2006. The role of BKCa channels in electrical signal encoding in the mammalian auditory periphery. *The Journal of Neuroscience* 26 (23): 6181–6189.
48. Pyott, S.J., et al. 2007. Cochlear function in mice lacking the BK channel alpha, beta1, or beta4 subunits. *The Journal of Biological Chemistry* 282 (5): 3312–3324.
49. Ricci, A.J., A.C. Crawford, and R. Fettiplace. 2003. Tonotopic variation in the conductance of the hair cell mechanotransducer channel. *Neuron* 40 (5): 983–990.
50. He, D.Z., S. Jia, and P. Dallos. 2004. Mechanoelectrical transduction of adult outer hair cells studied in a gerbil hemicochlea. *Nature* 429 (6993): 766–770.
51. Beurg, M., et al. 2006. A large-conductance calcium-selective mechanotransducer channel in mammalian cochlear hair cells. *The Journal of Neuroscience* 26 (43): 10992–11000.
52. Ricci, A.J., et al. 2005. The transduction channel filter in auditory hair cells. *The Journal of Neuroscience* 25 (34): 7831–7839.
53. Beurg, M., K.X. Kim, and R. Fettiplace. 2014. Conductance and block of hair-cell mechanotransducer channels in transmembrane channel-like protein mutants. *The Journal of General Physiology* 144 (1): 55–69.
54. Beurg, M., A.C. Goldring, and R. Fettiplace. 2015. The effects of Tmc1 Beethoven mutation on mechanotransducer channel function in cochlear hair cells. *The Journal of General Physiology* 146 (3): 233–243.
55. Beurg, M., et al. 2015. Subunit determination of the conductance of hair-cell mechanotransducer channels. *Proceedings of the National Academy of Sciences of the United States of America* 112 (5): 1589–1594.
56. Holt, J.R., et al. 1999. Functional expression of exogenous proteins in mammalian sensory hair cells infected with adenoviral vectors. *Journal of Neurophysiology* 81 (4): 1881–1888.
57. Jero, J., et al. 2001. Cochlear gene delivery through an intact round window membrane in mouse. *Human Gene Therapy* 12 (5): 539–548.

58. Gubbels, S.P., et al. 2008. Functional auditory hair cells produced in the mammalian cochlea by in utero gene transfer. *Nature* 455 (7212): 537–541.
59. Akil, O., et al. 2012. Restoration of hearing in the VGLUT3 knockout mouse using virally mediated gene therapy. *Neuron* 75 (2): 283–293.
60. Chen, W., et al. 2012. Restoration of auditory evoked responses by human ES-cell-derived otic progenitors. *Nature* 490 (7419): 278–282.
61. Mizutari, K., et al. 2013. Notch inhibition induces cochlear hair cell regeneration and recovery of hearing after acoustic trauma. *Neuron* 77 (1): 58–69.
62. Lentz, J.J., et al. 2013. Rescue of hearing and vestibular function by antisense oligonucleotides in a mouse model of human deafness. *Nature Medicine* 19 (3): 345–350.
63. Pearson, R.A., et al. 2012. Restoration of vision after transplantation of photoreceptors. *Nature* 485 (7396): 99–103.
64. Polosukhina, A., et al. 2012. Photochemical restoration of visual responses in blind mice. *Neuron* 75 (2): 271–282.
65. Nelson, C.E., et al. 2016. In vivo genome editing improves muscle function in a mouse model of Duchenne muscular dystrophy. *Science* 351 (6271): 403–407.
66. Long, C., et al. 2016. Postnatal genome editing partially restores dystrophin expression in a mouse model of muscular dystrophy. *Science* 351 (6271): 400–403.
67. Tabebordbar, M., et al. 2016. In vivo gene editing in dystrophic mouse muscle and muscle stem cells. *Science* 351 (6271): 407–411.
68. Zuris, J.A., et al. 2015. Cationic lipid-mediated delivery of proteins enables efficient protein-based genome editing in vitro and in vivo. *Nature Biotechnology* 33 (1): 73–80.
69. Landegger, L.D., et al. 2017. A synthetic AAV vector enables safe and efficient gene transfer to the mammalian inner ear. *Nature Biotechnology* 35 (3): 280–284.
70. Gao, X., et al. 2017. Treatment of autosomal dominant hearing loss by in vivo delivery of genome editing agents. *Nature* 553: 217–221.
71. Pepermans, E., et al. 2014. The CD2 isoform of protocadherin-15 is an essential component of the tip-link complex in mature auditory hair cells. *EMBO Molecular Medicine* 6 (7): 984–992.

Chapter 6
Summary: Mechanotransduction and Beyond

Wei Xiong

Keywords Tethering model · Asymmetry · Tonotopy · Molecular assembly · Pore-forming subunit

In the previous chapters, we discussed the molecular, biophysical, and physiological roles of mechanotransduction (MET) in auditory hair cells and cochlear function, in addition to a minimal but necessary background on anatomy and history of cochlea study. As the final summary in this chapter, we will extend our discussion to open questions and current dilemma still in debate when summarizing consensus we have had.

The core element is the MET channel whose functionality recruits more than a dozen of proteins to form a complex. It defines the celebrity role of the hair cell in the inner ear and maximizes the function of the cochlea that is designed to a sophisticated structure. As a channel with elusive identity yet, it performs the task faithfully but no lack of variation.

- MET properties of the hair cell are quite unique due to its molecular composition that is so complicated and not similar to any currently known mechanical sensation system. Thirteen components have been characterized to participate MET with partially known molecular function. A "tethering model" has been proposed to describe its working form.
- Similar with other mechanical sensation, adaptation is widely used in the hair-cell MET, which is not identical to inactivation in terms of channel operation. Nevertheless, forms of fast adaptation and slow adaptation have been described on the hair cell. Fast adaptation is believed to be an intrinsic property of the MET channel that is mainly mediated by the entering calcium, while slow adaptation is considered to represent tension of the tip-link complex including its intracellular components.

W. Xiong (✉)
School of Life Sciences, IDG/McGovern Institute for Brain Research, Tsinghua University, Beijing, China
e-mail: wei_xiong@mail.tsinghua.edu.cn

© The Author(s) 2018   59
W. Xiong, Z. Xu (eds.), *Mechanotransduction of the Hair Cell*,
SpringerBriefs in Biochemistry and Molecular Biology,
https://doi.org/10.1007/978-981-10-8557-4_6

- Asymmetry is a typical trait in hair cells at either cellular or molecular levels. Hair bundles possess polarity towards the same direction of shear stress. Organization of stereocilia is in a staircase fashion with molecular linkages for its stabilization. Tip links are specialized linkages that gate the MET channel in addition to bridge the tops of stereocilia connection. The tip links are composed of two different cadherin molecules as the upper and lower parts. The MET channels localize only at the lower tip-link ending.
- Tonotopic variation is not only represented by gradient manifestations such as cochlear structure, basilar membrane properties, length of the OHC somas and hair bundles, and width of the stereocilium array but also biophysical properties of the MET channels.

However, there are still several answers elusive in terms of MET of the hair cell.

- What on earth is the identity of pore-forming subunit of the transducer channel? Researchers have utilized most of the state-of-art approaches to clone/purify candidates. However, the channel identity is always ambiguous. TMC1/2 has been proposed as the potential candidate except that it is difficult to reconstruct MET function ectopically in cell line expression system. There are two possibilities. If TMC1/2 forms the channel, an experiment needed to be done is to reconstruct the mechanotransducer complex on the cell surface. This is the gold standard to verify a channel, and of course the lipid bilayer system is also the ultimate validation. If TMC1/2 is not the essential core, further screen is still needed. Due to lack of a hair-cell-like cell line system, currently there is no efficient way to screen the channel. From mouse genetics and human genetics study, many important proteins were characterized but still missing the core. An ex vivo system might be helpful to solve the question.
- Why does the transducer complex need so many components? What is the goal for this assembly? What is the mechanism of altered property of the transducer channels that vary along the cochlear coil? Is this alteration correlated with the complex assembly? Does the variance of channel properties correlate with cochlear mechanics and tuning?
- Bundle morphology is unique by forming a staircase shape. However, it is still unclear what decides the height of each stereocilium during its elongation? Does MET contribute to the determination of the graded lengths of stereocilia?

Just like lacking of parts in a jigsaw puzzle game, we cannot get a whole picture yet on the molecular and working mechanisms of MET of the hair cell. Currently, the technical advance in biology is undoubtedly advancing the research on auditory transduction. Here are some examples. Cryo-electron tomography could gain better understanding of the bundle structure at cellular level, even at molecular level. Spatiotemporal transcriptome analysis at single-cell level may provide unprecedented opportunity to elucidate molecules actively contributing to auditory transduction. It is predictable that in the near future emerging evidence from anatomical, biophysical, molecular, and physiological studies will provide us a clear diagram to explain how the hair cell works to fulfil the acoustic decoding.

Printed in the United States
By Bookmasters